Iris Eisenbach

English for Materials Science and Engineering

Iris Eisenbach

English for Materials Science and Engineering

Exercises, Grammar, Case Studies

VIEWEG+
TEUBNER

Bibliographic information published by the Deutsche Nationalbibliothek
The Deutsche Nationalbibliothek lists this publication in the Deutsche Nationalbibliografie;
detailed bibliographic data are available in the Internet at http://dnb.d-nb.de.

Iris Eisenbach has extensive experience in teaching all levels of English to speakers of other languages and for a wide range of educational and professional purposes. The author graduated in English and French from the University of Mainz and from the Teacher Training College (Studienseminar) in Wiesbaden, (both Germany) with the Second State Examination. After teaching foreign languages to students at different levels for some years, she got tenure as a civil servant (Oberstudienrätin). Iris Eisenbach has spent the past 20 years concentrating on teaching English to students in university settings. Presently she is working as a university language instructor at the English and German Departments of the Language Center of the University of Stuttgart, Germany.

1st Edition 2011

All rights reserved
© Vieweg+Teubner Verlag | Springer Fachmedien Wiesbaden GmbH 2011

Editorial Office: Imke Zander | Thomas Zipsner

Vieweg+Teubner Verlag is a brand of Springer Fachmedien.
Springer Fachmedien is part of Springer Science+Business Media.
www.viewegteubner.de

No part of this publication may be reproduced, stored in a retrieval system or transmitted, in any form or by any means, electronic, mechanical, photocopying, recording, or otherwise, without the prior written permission of the copyright holder.

Registered and/or industrial names, trade names, trade descriptions etc. cited in this publication are part of the law for trade-mark protection and may not be used free in any form or by any means even if this is not specifically marked.

Cover design: KünkelLopka Medienentwicklung, Heidelberg
Layout: Stefan Kreickenbaum, Wiesbaden
Pictures: Graphik & Text Studio, Dr. Wolfgang Zettlmeier, Barbing; Stefan Kreickenbaum, Wiesbaden
Printed on acid-free paper
Printed in Germany

ISBN 978-3-8348-0957-5

Introduction

This textbook is intended for students of materials science, of different branches of engineering and of related disciplines who need to re-activate their English language skills. Using authentic materials and figures selected from scientific texts, students will improve their reading, writing and speaking skills in a context relevant to their specialist studies. This work does not attempt to teach the subject of materials science.

In addition to covering linguistic features specific to scientific and technical purposes, this book also presents review and practice activities in common problem areas of general English usage. The material for the textbook has been developed and tested in classes at the English Department of the University of Stuttgart over several semesters, and it addresses most of the problems English-language learners confront.

Students' feedback has been incorporated into the textbook; the author gratefully acknowledges these contributions, which make the book useful for successful teaching and self-study purposes.

Since the book is designed as both textbook and workbook, it is suitable for classroom use and for self-study. It contains extensive monolingual glossaries, tasks, grammar reviews and word studies directly related to the texts and figures. Solutions are offered in the back of the book.

The textbook offers sufficient material for a one-semester language class of about 14 sessions. Subjects, grammar reviews and word studies can also be studied independently.

Acknowledgements

This book would never have been written without the support of the Materials Research Laboratory (MRL) of the University of California, Santa Barbara, where I was accompanying my husband, Professor Claus D. Eisenbach, in 2007–2008. I am very grateful to the MRL for kindly offering me the use of the visiting scholar's office and for providing equipment and support.

The MRL also made it possible for me to attend classes by two excellent researchers and dedicated teachers, Professor Ram Seshadri and Professor Susanne Stemmer. Professor Seshadri in particular introduced me to the field of materials science and directed me to my most valuable source, *Materials Science and Engineering: An Introduction*, by William D. Callister Jr.

I am also indebted to my husband who was a constant source of knowledge and expertise and who read and commented on the manuscript. Special thanks to my good friend Pamela Lavigne, whose experience in TESOL (Teaching English to Speakers of Other Languages) and in editing were of great help. I am likewise grateful to the editors of "Lektorat Maschinenbau" at Vieweg+Teubner for their technical assistance.

Stuttgart, Autumn 2010   Iris Eisenbach

Table of contents

Chapter 1 Introduction 1
    1.1   Historical Background 1
    1.2   Grammar: Simple Past versus Present Perfect 3
    1.3   Materials Science versus Materials Engineering 5
    1.4   Selection of Materials 6
    1.5   Some Phrases for Academic Presentations 7
    1.6   Case Study: The Turbofan Aero Engine 8
    1.7   Some Abbreviations for Academic Purposes 10

Chapter 2 Characteristics of Materials 12
    2.1   Structure 12
    2.2   Some Phrases for Academic Writing 13
    2.3   Case Study: The Gecko 15
    2.4   Property 16
    2.5   Some Phrases for Describing Figures, Diagrams and for Reading Formulas 19
    2.6   Grammar: Comparison 20
    2.7   Processing and Performance 21
    2.8   Classification of Materials 23
    2.9   Grammar: Verbs, Adjectives, and Nouns followed by Prepositions 24

Chapter 3 Metals 25
    3.1   Introduction 25
    3.2   Mechanical Properties of Metals 27
    3.3   Important Properties for Manufacturing 29
    3.4   Metal Alloys 30
    3.5   Case Study: Euro Coins 32
    3.6   Grammar: Adverbs I 34
    3.7   Case Study: The Titanic 35
    3.8   Grammar: The Passive Voice 36
    3.9   Case Study: The Steel-Making Process 38

Chapter 4 Ceramics 40
    4.1   Introduction 40
    4.2   Structure of Ceramics 41
    4.3   Word Formation: Suffixes in Verbs, Nouns and Adjectives 41
    4.4   Properties of Ceramics 43
    4.5   Case Study: Optical Fibers versus Copper Cables 44
    4.6   Grammar: Adverbs II 46
    4.7   Case Study: Pyrocerams 46
    4.8   Case Study: Spheres Transporting Vaccines 48
    4.9   Useful Expressions for Shapes and Solids 49

Chapter 5 Polymers ... 51

- 5.1 Introduction ... 51
- 5.2 Word Formation: The Suffix -able/-ible ... 52
- 5.3 Properties of Polymers ... 53
- 5.4 Case Study: Common Objects Made of Polymers ... 54
- 5.5 Case Study: Ubiquitous Plastics ... 55
- 5.6 Grammar: Reported Speech (Indirect Speech) ... 57
- 5.7 Polymer Processing ... 59
- 5.8 Case Study: Different Containers for Carbonated Beverages ... 61

Chapter 6 Composites ... 63

- 6.1 Introduction ... 63
- 6.2 Case Study: Snow Ski ... 64
- 6.3 Grammar: Gerund (-ing Form) ... 66
- 6.4 Case Study: Carbon Fiber Reinforced Polymer (CFRP) ... 69
- 6.5 Word Formation: Prefixes ... 70

Chapter 7 Advanced Materials ... 73

- 7.1 Introduction ... 73
- 7.2 Semiconductors ... 75
- 7.3 Case Study: Integrated Circuits ... 76
- 7.4 Grammar: Subordinate Clauses ... 77
- 7.5 Smart Materials ... 78
- 7.6 Nanotechnology ... 80
- 7.7 Case Study: Carbon Nanotubes ... 80
- 7.8 Grammar: Modal Auxiliaries ... 82

KEY ... 84

Credits/Selected Reference List ... 104

Glossary ... 106

Chapter 1 Introduction

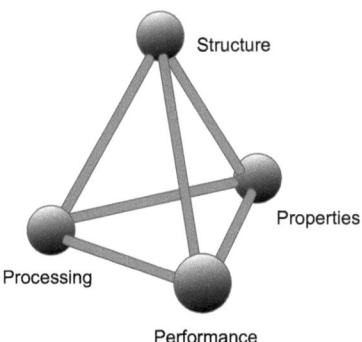

Figure 1: Materials science tetrahedron [wikipedia]

1.1 Historical Background

Task 1. *Work with a partner. Fill the gaps in the text with words from the box in their correct form.*

alloy; characteristic; communication; clay; crystal; heat; housing; manipulate; metal; pottery; property (2); skin; specimen; substance; structure; technological; wood

Materials used in food, clothing, ..., transportation, recreation and .... influence virtually every segment of our everyday lives.
Historically, materials have played a major role in the development of societies, whose advancement depended on their access to materials and on their ability to produce and .... them. In fact, historians named civilizations by the level of their materials development, e.g. the Stone Age (beginning around 2.5 million BC), the Bronze Age (3500 BC), and the Iron Age (1000 BC). The earliest humans had access to only a very limited number of materials, those that occur naturally, e.g. ...., ... and .... With time they discovered techniques for producing materials that had properties superior to those of the natural ones; these new materials included .... and various .... Furthermore, early humans discovered that the properties of a material could be altered

by .... treatments, e.g. to soften metals, and by adding other .... to produce a new material, e.g. by melting copper, then mixing it with tin to form bronze which could be regarded as the first ....

Until recently, selecting a material involved choosing from a number of familiar materials the one most appropriate for the intended application by virtue of its characteristics but without knowing much about its structure. Only in the 19th century did scientists begin to understand the relationships between the structural elements of materials and their .... In 1864 the Englishman Henry Sorby first showed the microstructure of a metal when he developed a technique for *etching* the surface layer of a polished metal .... by a chemical reaction. He used a light reflecting microscope to show that the material consisted of small .... which reflected the light in different ways because they were oriented in different directions. The crystals were well fitted together and joined along *grain boundaries*.

Modern techniques such as x-ray diffraction, transmittance electron microscopy (TEM) and scanning electron microscopy (SEM) make possible to see further into the .... of materials, which leads to a better understanding of their characteristics and promotes intentional alteration and improvement of their .... By now more than 50,000 materials with specialized .... have been developed and are available to the engineer, who has to choose the one best suited to serve the given purpose. Since much of what can be done .... is limited by the available materials, engineers must constantly develop new materials with improved properties.

(from Callister, modified and abridged)

Glossary

to etch	to cut into a surface, e.g. glass, using an *acid*
acid	a chemical, usually a sour liquid, that contains hydrogen with a pH of less than 7
grain boundary	a line separating differently oriented crystals in a polycrystal

Task 2. *Different verbs in English can be used to describe the action of changing, such as adjust; alter; change; modify; transform; vary. Refer to a dictionary or thesaurus, then list the differences in usage and meaning.*

Task 3. *Give a short explanation for x-ray diffraction, TEM and SEM.*

1.2 Grammar: Simple Past versus Present Perfect

Scientific and technical texts in English frequently use the present tense, since in most cases they state facts. Sometimes, the present perfect and simple past have to be used, as the text about the historical development of materials science shows.

Formation of the Simple Past

Use the so-called second form of the verb

write – *wrote* – written

She *wrote* the second proposal last month.

Formation of the Present Perfect

Use have/has + the third form of the verb (the past participle).

write – wrote – *written*

She *has* just *written* the second proposal.

> **Use of the Simple Past**
>
> Use the simple past for actions in the past that have **no** connection to the present and when the **time** of the past action **is important** or **shown.**
>
> Signal words are yesterday, last Thursday, two weeks ago, in November 1989

> **Use of the Present Perfect**
>
> Use the present perfect for actions in the past **with** a connection to the present and when the **time** of the past actions **is not important.**
>
> Use the present perfect for recently completed actions and actions beginning in the past and continuing in the present.
>
> Signal words are: just, never, ever, yet, already, recently, since, for, so far, up to now

Task 1. Work in a group. Revise English irregular verbs, by using a table, e.g. from a dictionary or English grammar book. Take turns eliciting the correct forms from members of your group.

Task 2. Work with a partner. Fill the gaps in the sentences with the verbs in their correct tense (present perfect or simple past).

Materials .... (always play) a major role in the development of societies.

Civilizations .... (designate) by the level of their materials development.

The earliest humans .... (have) access to only a very limited number of materials.

The microstructure of a metal .... (be) first revealed in 1864 by the Englishman Henry Sorby who .... (develop) a technique for etching the surface layer of a polished metal.

Modern techniques such as x-ray diffraction, transmission electron microscopy (TEM) and scanning electron microscopy (SEM) .... (make) it possible to better understand their characteristics.

By now, more than 50,000 materials .... (develop).

Materials scientists .... (long envy) the *resilience* of certain naturally occurring materials.

Past efforts to reproduce the architecture of, e.g. a shell .... (not be successful).

To copy the microstructure of the shell, the researchers .... (mix) water with finely ground ceramic powder and polymer *binders*.

Glossary

resilience, *n* resilient, *adj*	elasticity; property of a material to resume its original shape/position after being bent/stretched/compressed
binder	a polymeric material used as *matrix* in which particles are evenly distributed
matrix	a substance in which another substance is contained

n = noun   *adj* = adjective   *v* = verb

1.3 Materials Science versus Materials Engineering

The discipline of materials science and engineering includes two main tasks.
Materials scientists examine the structure-properties relationships of materials and develop or *synthesize* new materials.

Materials engineers design the structure of a material to produce a *predetermined* set of properties on the basis of structure-property relationships. They create new products or systems using existing materials and/or develop techniques for processing materials.

Most graduates in materials programs are trained to be both materials scientists and materials engineers.

(from Callister, modified and abridged)

Glossary

to synthesize, synthesis, *n*	to produce a substance by chemical or biological reactions
predetermined	decided beforehand

Task 1. *Read the text above. Then decide whether the statements are true or false. Rewrite the false statements if necessary.*

Materials scientists do research on finished materials.

....

New products are based on new materials only.

....

Materials science can be subdivided because different approaches to materials are employed.

....

Materials engineers investigate the correlation between structure and property.

....

1.4 Selection of Materials

Selecting the right material from the many thousands that are available poses a serious problem. The decision can be based on several criteria. The in-service conditions must be characterized, for these will dictate the properties required of the material. A material does not always have the maximum or ideal combination of properties. Thus, it may be necessary to trade off one characteristic for another.

The classic example includes *strength* and *ductility*. Normally, a material having a high strength will have only a limited ductility. A second selection consideration is any deterioration of material properties that may occur during service operation.

For example, significant reductions in mechanical strength may result from exposure to elevated temperatures or *corrosive* environments. If a compromise concerning desired in-service properties cannot be reached, new materials have to be developed.

Probably the most important consideration is that of economics. A material may be found that has the ideal set of properties but is extremely expensive. Some compromise is inevitable. The cost of a finished piece also includes any cost occurring during fabrication to produce the desired shape. For example: *commodity* plastics like polyethylene or polypropylene cost about $ 0.50/*lb*, whereas engineering *resins* or Nylon cost $ 1,000/lb.

(from Callister, modified and abridged)

Glossary

strength	the power to resist stress or strain; the maximum load, i.e. the applied force, a *ductile* material can withstand without permanent deformation
ductility, *n* ductile, *adj*	a material's ability to suffer measurable *plastic deformation* before fracture
plastic deformation	a non-reversible type of deformation, i.e. the material will not return to its original shape
corrosive, *n, adj* to corrode, corrosion	a corroding substance, e.g. an acid
commodity	article of trade
lb	pound, 453.592 grams
resin	a natural substance, e.g. amber, or a synthetic *compound*, which begins in a highly *viscous* state and hardens when treated
compound	a pure, macroscopically homogeneous substance consisting of atoms/ions of two/more different elements that cannot be separated by physical means
viscous, *adj* viscosity, *n*	having a relatively high resistance to flow

Task 1. *Explain the grammatical use of the term prohibitively in the sentence below.*

A material may be found that has the ideal set of properties but is prohibitively expensive.

Task 2. *Write short answers to the questions.*

What are necessary steps when considering a material for a certain application?

....
....
....
....

Which trade-offs are unavoidable when choosing a particular material?

....
....
....
....

1.5 Some Phrases for Academic Presentations

Introduction (after greeting the audience and introducing yourself or being introduced)
The subject/topic of my presentation today will be …
Today I would like to present recent result of our research on …
What I want to focus on today is …

Outlining the structure of the presentation
I will address the following three aspects of …
My presentation will be organized as can be seen from the following slide.
I will start with a study of … . Next, important discoveries in the field of … will be introduced.
Finally, recent findings of … will be discussed.

Introducing a new point or section
Having discussed …, I will now turn to …
Let's now address another aspect.

Referring to visual aids
As can be seen from the next slide/diagram/table …
This graph shows the dependency of … versus …
The following table gives typical values of …
In this graph we have plotted … with …

Concluding/summarizing

Wrapping up …

To summarize/sum up/conclude …

Inviting questions

Please don't hesitate to interrupt my talk when questions occur.

I'd like to thank you for your attention.

I'll be happy/pleased to answer questions now.

Dealing with questions

I cannot answer this question right now, but I'll check and get back to you.

Perhaps this question can be answered by again referring to/looking at table …

1.6 Case Study: The Turbofan Aero Engine

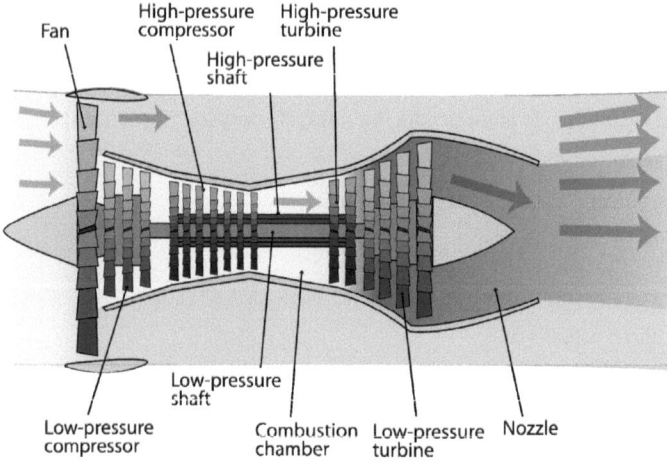

Figure 2: Cross-section of a turbofan aero engine [wikipedia]

Task 1. *Work with a partner. Study the following notes. Then refer to* 1.5 Phrases for Academic Presentation *and give a short presentation about the subject.*

In the turbofan aero engine, which is used to power large planes, air is propelled past and into the engine by the turbofan, providing aerodynamic *thrust*. The air is further compressed by compressor blades, then mixed with fuel and burnt in the *combustion* chamber. The expanding gases drive the turbine blades, which provide power to the turbofan and the compressor blades, and finally pass out of the rear of the engine, adding to the thrust.

1.6 Case Study: The Turbofan Aero Engine

Two kinds of materials were considered:

Metal, a titanium *alloy*

material's properties and in-service requirements:

Young's Modulus, yield strength, fracture toughness sufficiently good

high *density* (the heavier the engine, the less payload can be carried)

resistance to *fatigue* (due to rapidly varying loads)

resistance to surface wear (striking water drops, large birds)

resistance to corrosion (salt sprays from ocean entering the engine)

Composite, carbon-fiber reinforced polymer (CFRP)

material's properties and in-service requirements:

low density (half of that of titanium)

low weight

low toughness (potential deformation of blade by bird strike)

The problem posed by choosing CFRP for a blade can be overcome by cladding, which means giving the CFRP a metallic leading edge.

(from Ashby/Jones, modified and abridged)

Glossary

thrust	a forward directed force
combustion	the process of burning; here of fuel
alloy	a metallic substance that is composed of two or more elements which keep the same crystal structure in the alloy
Young's Modulus	elastic modulus (E), a material's property that relates *strain* (ε, epsilon) to applied *stress* (σ, sigma)
strain	the response of a material when *tensile stress* is applied
tensile stress	a force tending to tear a material apart
stress, *n*	the force applied to a material per unit area; (σ, sigma = F/A or lb/in^2)
in	inch, 2.54 cm
yield strength	the point at which a material starts to deform permanently
fracture toughness	the measure of a material's resistance to fracture when a *crack* occurs
crack, *n, v*	a break, fissure on a surface
density	mass per volume
fatigue	the weakening/failure of a material resulting from prolonged stress

1.7 Some Abbreviations for Academic Purposes

Task 1. Add your notes in the column on the right.

AC	alternating current	
approx., ca.	approximate(ly)	
AT	air temperature	
at. no.	atomic number	
at. wt.	atomic weight	
avg.	average	
b.p.	boiling point	
c., cu., cub.	cubic	
cath.	cathode	
cc	cubic centimetre(s)	
cf. (conferre)	confer, compare	
C. of C.	coefficient of correlation	
co.	column	
cont(d).	continue(d), contain(ed)	
ctr.	center	
DC	direct current	
Dept.	department	
dup.	duplicate	
e.g. (exempli gratia)	for example	
esp.	especially	
est(d).	estimated	
etc. (et cetera)	and so on	
ex.	example	
f., ft.	foot, feet, frequency	
hor.	horizontal	
i.e. (id est)	that is	
in., ins.	inch(es)	

1.7 Some Abbreviations for Academic Purposes

incl.	including, included, inclusive	
kWh	kilowatt-hour(s)	
l., ll.	long, length, line, lines	
liq.	liquid	
max., min.	maximum, minimum	
mech.	mechanical	
misc.	miscellaneous	
mol wt.	molecular weight	
m.p.	melting point	
n.a.	not applicable	
NB, nb (nota bene)	note particularly	
No., no.	number	
ord.	ordinary, ordinal	
oz(s).	ounce(s)	
par.	parallel	
prev.	previous	
pt.	part	
qt.	quantity, quart	
resp.	respectively	
rpm	revolutions per minute	
stat.	statistics	
std.	standard	
syn.	synthetic	
tech.	technical(ly)	
vel.	velocity	
vs.	versus	
w/	with	
w/o	without	
yd(s).	yard(s)	

Chapter 2 Characteristics of Materials

2.1 Structure

The structure of a material is usually determined by the arrangement of its internal components. On an atomic level, structure includes the organization of atoms relative to one another. Subatomic structure involves electrons within individual atoms and interactions with their nuclei. Some of the important properties of solid materials depend on geometrical atomic arrangements as well as on the interactions that exist among atoms or molecules.

Various types of primary and secondary interatomic bonds hold together the atoms composing a solid.

The next larger structural area is of nanoscopic scale which comprises molecules formed by the bonding of atoms, and particles or structures formed by atomic or molecular organisation, all within 1 *nm* – 100 nm dimensions. Beyond nano scale are structures called microscopic, meaning that they can directly be observed using some kind of microscope. Finally, structural elements that may be viewed with the naked eye are called macroscopic.

(from Callister, modified and abridged)

Glossary

nm	nanometer (10^{-9} m)

Task 1. Work with a partner. Fill in the table with the different structural levels and their characteristics as described in the text.

structural level	characteristics

Task 2. *Choose the correct terms for the following definitions.*

A sufficiently stable, electrically neutral group of at least two units in a definite arrangement held together by strong chemical bonds. ....

The smallest particle characterizing an element ....

A fundamental subatomic particle, carrying a negative electric charge. ....

It makes up almost all the mass of an atom. ....

A positively charged subatomic particle. ....

An electrically neutral subatomic particle. ....

2.2 Some Phrases for Academic Writing

Introduction
In this paper/project/article we will focus on …
In our study, we have investigated …
Our primary objective is …

Making a generalization
It is well known that …
It is generally accepted that …

Making a precise statement
In particular
Particularly/especially/mainly/ more specifically

Quoting
According to/referring to …
As has been reported in … by …
Referring to earlier work of …

Introducing an example
e.g. …
if … is considered for example

Interpreting

The data could be interpreted in the following way …

These data infer that …

This points to the fact that …

Referring to data

As is shown in the table/chart/data/diagram/graph/plot/figure

Adding aspects

Furthermore our data show …

In addition … has to be considered

Expressing certainty

It is clear/obvious/certain/noticeable that …

An unequivocal result is that …

Expressing uncertainty

It is not yet clear whether …

However it is still uncertain/open if …

Emphasizing

It has to be emphasized/stressed that …

Summarizing

Our investigation has shown that …

To summarize/sum up our results …

Concluding

We come to the conclusion that …

Our further work will focus on …

Further studies/research on … will still be needed.

Detailed insights into … are still missing.

2.3 Case Study: The Gecko

Figure 3: The underside of a gecko and its feet [adapted from Seshadri]

Task 1. *Work with a partner. Fill the gaps in the text with words from the box in their correct form. Some terms are used more than once.*

| adhesion; *adhesive;* design; horizontal; mass; microscopic; molecule; *release; residue;* self-cleaning; sticky; surface; underside; vertical |

The photograph shows the .... of a gecko, a harmless tropical lizard, and its toes. Researchers worldwide are studying the animal's adhesive system. The scientists want to learn from nature how to .... dry adhesives such as geckos apply when moving their feet over smooth surfaces. The animals achieve high adhesion and friction forces required for rapid .... (running up walls) and inverted (running along the underside of .... surfaces) motion, since their .... feet will cling to virtually any surface. Yet they can easily and quickly release the sticky pads under their toes to make the next step. A gecko can support its body .... with a single toe, because it has an extremely large number of .... small ordered fiber bundles on each *toe pad*. When these fibrous structures come in contact with a surface, weak forces of attraction, i.e. van der Waals forces, are established between hair .... and molecules on the surface. The fact that these fibers are so small and so numerous explains why the animal grips .... so tightly. To .... its grip, the gecko simply curls up its toes and peels the fibers away from the surface. Another fascinating feature of gecko toe pads is that they are .... that is, dirt particles don't stick to them. Scientists are just beginning to understand the mechanism of .... for these tiny fibers, which may lead to the development of .... self-cleaning synthetics. Imagine *duct tape* that never looses its stickiness or bandages that never leave a sticky ....

(from Callister, modified and abridged)

Glossary

adhesive *n, adj,* to adhere, adhesion, *n*	a substance used for joining surfaces together, sticky
release, *v, n*	to let go
residue	the remainder of sth after removing a part
toe pad	a cushion-like flesh on the underside of animals' toes and feet
duct tape	an adhesive tape for sealing heating and air-conditioning ducts

2.4 Property

While in use, all materials are exposed to external stimuli that cause some kind of response. A property is a material characteristic that describes the kind and magnitude of response to a specific stimulus. For example, a specimen exposed to forces will experience deformation, or a metal surface that has been polished will reflect light. In general, definitions of property are made independent of material shape and size.

Virtually all important properties of solid materials may be grouped into six different categories:
- mechanical
- electrical
- thermal (including melting and *glass transition temperatures*)
- magnetic
- optical
- deteriorative

(from Callister, modified and abridged)

Glossary

glass transition temperature T_g	the temperature at which, upon cooling, a non-crystalline ceramic transforms from a supercooled liquid to a solid glass
supercooled	cooled to below a phase transition temperature without the occurrence of transformation

2.4 Property

Mechanical Properties relate deformation to an applied load or force; examples include *elastic modulus* and strength.

Glossary

elastic modulus (E)	or Young's Modulus, a material's property that relates strain (ε, epsilon) to applied stress (σ, sigma), cf. p. 9

Electrical Properties are, e.g. electrical *conductivity, resistivity* and *dielectric constant*. The stimulus is voltage or an electric field.

Glossary

conductivity	ability to transmit heat and/or electricity
resistivity	a material's ability to oppose the flow of an electric current
dielectric constant	a measure of a material's ability to resist the formation of an electric field within it

Thermal Properties of solids can be described by heat capacity and thermal conductivity.

Poor thermal conductivity is responsible for the fact that space shuttle *tiles* containing amorphous, porous silica (SiO_2) can be held at the corners, even when glowing at 1000 °C.

Glossary

tile	a flat, square piece of material

Task 1. Work with a partner. Refer to the texts, then answer the questions.

What is a material's property?

....

Do mechanical properties deal with deformation?

....

How can the thermal behavior of solids be characterized?

....

Magnetic Properties demonstrate a material's response to the application of a magnetic field.

Optical Properties are a material's response to electromagnetic or visible light. The index of *refraction* and *reflectivity* are representative optical properties.

Glossary

refraction	the bending of a light beam upon passing from one medium into another
reflectivity	the ability to reflect, i.e. to change the direction of a light beam at the interface between two media

Deteriorative Properties relate to the chemical reactivity of materials. The chemical reactivity, e.g. corrosion, of a material such as an alloy, can be reduced by heat treating the alloy prior to exposure in salt water. Heat treatment changes the inner structure of the alloy. Thus crack propagation leading to mechanical failure can be delayed.

Glossary

propagation	the process of spreading to a larger area

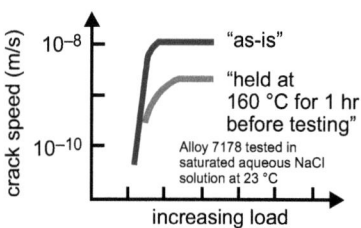

Figure 4:
Crack propagation and load [adapted from Seshadri]

Task 2. *Refer to 2.5 Some Phrases for Describing Figures, Diagrams and Reading Formulas and write a short paragraph for the plot in the figure above, describing what is shown.*

The graph in the figure above shows

....

....

....

....

....

....

2.5 Some Phrases for Describing Figures, Diagrams and for Reading Formulas

Graph/Diagram

the graph/diagram/figure represents …
it shows a value for …
it shows the relationship between …
the curve shows a steep *slope*, a peak, a trough
the curve rises steeply/flattens out/drops/extrapolates to zero

Plot

to plot points on/along an axis
to plot/make a plot … versus … for …
x is plotted as a function of y

Coordinate System

abscissa (x-axis) and ordinate (y-axis)
the coordinate system shows the frequency of … in relation to/per …

Angle

parallel; perpendicular; horizontal to
right angle (90°)
acute angle (smaller than 90°)
obtuse angle (larger than 90°)
straight angle (180°)

Mathematics

to apply a law
to equal, to be equal to
to calculate/compute
to determine/assume/substitute a value
to *derive* an equation
in a fraction, there are numerator and divisor (denominator)

Glossary

slope	a line that moves away from horizontal
to derive	to deduce; to obtain (a function) by differentiation

Task 1. *Complete the table.*

10,000	is read ten thousand
0.28	is read …
$1/4$	
$1/12$	one over twelve
$6\,3/5$	
x^2	
x^3	
x^{-4}	
$\sqrt{4}$	
$\sqrt[3]{a}$	
$1/x$	
a_n	
$^n a$	

Glossary

slope	a line that moves away from horizontal
to derive	to deduce; to obtain (a function) by differentiation

2.6 Grammar: Comparison

Comparing Two or more Things in English

Add **-er** and **–est** to adjectives with one syllable
　　strong – stronger – strongest
to adjectives with two syllables and ending with **-y**
　　oily – oilier – oiliest
Use **more** and **most** for adjectives with more than two syllables and not ending with -y
　　resistant – more resistant – most resistant.
for adverbs
Polyethylene is more frequently produced than poly(tetrafluoro ethylene).

2.7 Processing and Performance

Task 1. Fill the gaps in the table with the correct forms.

Irregular Forms:

good	....	....	
bad	....	....	
far	....	....	(when referring to distance)
far	....	....	(when referring to extent/degree)
little	....	....	(when referring to amount)
little	....	....	(when referring to size)
much/many	....	....	

Use **as ... as** when comparing items of the same characteristics.
> Physics is as interesting as chemistry.

Use **not as (so) ... as** when comparing items of dissimilar characteristics.
> Polymers are not as brittle as ceramics.

Alternatively use **-er / more ... than**.
> Some alloys are easier to process than others.

2.7 Processing and Performance

In addition to structure and properties, materials differ in terms of processing and performance. Processing determines structure and structure affects property. Last, property influences performance.

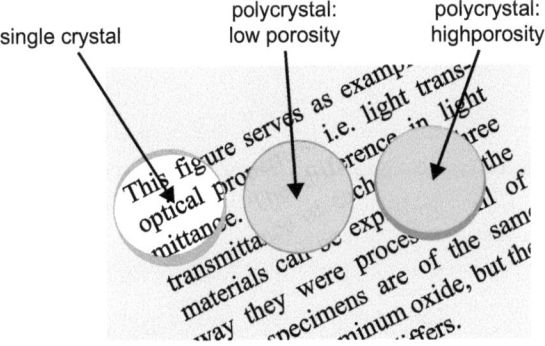

Figure 5: Crystallinity and light transmittance

This figure serves as example for optical properties, i.e. light transmittance. The difference in light transmittance of each of the three materials can be explained by the way they were processed. All of these specimens are of the same material, aluminum oxide, but their crystal structure differs.

Task 1. *Work with a partner. Complete the short paragraph for the figure above, explaining the difference in optical properties.*

Figure 5 illustrates the relationship among processing, structure, properties and performance. The photograph shows three thin disk specimens of the same material, ..., placed over .... The optical properties (i.e. the light transmittance) of each of the three materials are different. The one on the left ...., i.e. virtually all of the light reflected from the printed page passes through it. The disk .... translucent, meaning that some of this .... through the disk. The disk on the right is ...., i.e. none of the .... passes through. Optical properties are a consequence of .... of these materials which result from the way the materials were processed. The leftmost one is a .... which causes its .... The polycrystal in the center is composed of numerous small crystals that are all connected, the *boundaries* between these small crystals *scatter* a portion of the ...., so this material is optically translucent. The specimen on the right is not only composed of many small interconnected crystals but also of many very small pores. These pores also effectively scatter the reflected light and make this material opaque.

(from Callister, modified and abridged)

Glossary

boundary	the interface separating two neighboring regions having different crystallographic orientation
to scatter	to distribute in all directions

2.8 Classification of Materials

Solid materials can be grouped into three basic classifications:

      metals, ceramics and polymers.

This classification is based primarily on chemical makeup and atomic as well as molecular structure. Most materials fall into one distinct grouping, although there are some intermediates. More engineering components are made of metals and alloys than of any other class of solid. But increasingly, polymers are replacing metals, because they offer a combination of properties more attractive to designers.

New ceramics are developed worldwide, which will permit materials engineers to devise more efficient heat engines and lower friction *bearings*. Ceramics have been found that become superconducting (showing electrical conductivity with very limited resistance) at extremely low temperatures (about 100 K, approximately minus 170 °C). If this phenomenon is ever achieved at *ambient temperature*, it may increase the use of ceramics and revolutionize electronics.

The best properties of materials can be combined to make composites which often combine two or more materials from these three basic classes. In high-technology applications, a new classification called advanced or smart materials emerges. These materials are semiconductors, biocompatible materials, and nano-engineered materials.

Natural materials like wood or leather should also be mentioned, since they offer properties that, even with the innovations of today's materials scientists, are hard to beat.

(from Callister and Ashby/Jones, modified and abridged)

Glossary

bearing	a device to reduce friction between a rotating staff and a part that is not moving
ambient temperature	the temperature of the air above the ground in a particular place; usually room temperature, around 20 – 25 °C

Task 1. *Read the text then decide whether the statements are true or false.*
Rewrite the false statements if necessary.

Polymers belong to a distinct material group.

....

Ceramics will increasingly be used for applications in electronics because of their hardness.

....

Man-made materials are superior to natural materials.

....

2.9 Grammar: Verbs, Adjectives, and Nouns followed by Prepositions

The texts above contain verbs, adjectives, and nouns that are followed by prepositions. Learning to use the correct preposition following a verb, adjective or noun can be challenging; particularly when the preposition differs from, e.g. German usage.

> to depend **on** – abhängen **von**.

Below are some examples taken from the texts you have worked with so far.

Task 1. Work with a partner. Add the correct prepositions to the terms. Give examples with collocations, i.e. two or more words often used together.

Verbs

to expose to *materials that are exposed to external stimuli*

to rely

....

to trade

....

to relate

....

Adjectives/ Participles

transparent

....

based

....

composed

....

according

....

Nouns

in response

....

decrease

....

in reference to

....

Chapter 3 Metals

3.1 Introduction

Metallic materials have large numbers of non-localized electrons; i.e. these electrons are not bound to particular atoms. Many properties of metals are directly attributable to these electrons, often referred to as electron gas, cloud or sea.

Task 1. *Work with a partner. Study the following notes. Then refer to the 2.2 Some Phrases for Academic Writing and write an introductory text about metals, adding details you know.*

Mechanical Properties

relatively *dense,* stiff and strong, ductile, resistant to fracture
hard and solid at ambient temperature,
except for: sodium (soft), mercury (liquid at room temperature)

Conductivity

very good conductors of electricity and heat
e.g. copper, iron (conduct heat better than stainless steel)

Optical Properties

opaque, colored
lustrous appearance of metal surface when polished, but
dull appearance after oxidization of surface by contact and reaction with air

Magnetic Property

most metals non-magnetic (including many steels)
some metals magnetic, e.g. iron, cobalt, nickel

Application

widespread applications *(add examples of your own)*
e.g. in construction, plumbing, electrical and mechanical engineering

Processing

molding, casting, plastic deforming, cutting, joining, etc. *(add examples)*

(from Callister, modified and abridged)

Glossary

dense, density, *n*	referring to mass per volume
lustrous, luster, *n*	shining brightly and gently

Task 2. *Work in a group. Add the chemical symbols of the metals and list what you know about them. Refer to the metal's properties and applications, as shown in the example.*

iron, Fe   a lustrous, malleable, ductile, magnetic or magnetizable metallic element occurring in minerals; rusts easily; used to make steel and other alloys, important in construction and manufacturing

copper ....

....

nickel ....

....

mercury ....

....

sodium ....

....

zinc ....

....

aluminum ....

....

gold ....

....

lead ....

....

tin ....

....

3.2 Mechanical Properties of Metals

Bend Strength

Fracturing, e.g. a *rod* of brittle material, can be done by fixing it tightly at both ends and applying a force upwards at two central points. Fracture will appear almost *perpendicular to* the length of the rod. This is one way of measuring the bend strength of material.

Shear Strength

Breaking the rod by fixing it at one end and twisting the other end, applying shear load or stress (τ, tau), will result in fracture that occurs at an oblique angle to the length of the rod.

Stress (σ, sigma) is the ratio of a force F to the area A on which the force acts:

$\sigma = F/A = lb/in^2$ (lb meaning 453.592 grams, in meaning inch).

Shear strength is important for rods of material that rotate like rotating *axles* in machinery which sometimes fail this way.

Tensile Strength

Most metals show macroscopically noticeable stretching. Brittle materials, like ceramics, show very little plastic, i.e. permanent deformation, before they fail.

Materials with high tensile strength, like plastic and rubber, will stretch to several times their original length before they break.

Glossary

rod	a thin, straight piece/bar, e.g. of metal, often having a particular function
perpendicular to	forming an angle of 90° with another line/surface
axle	a supporting shaft on which wheels turn

Task 1. Explain the testing of tensile strength in a few words with the help of Figure 6 below.

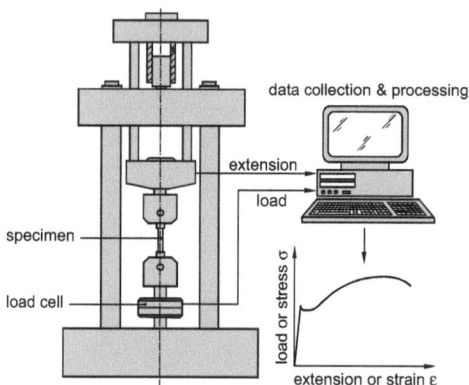

Figure 6:
Testing tensile strength [V. Läpple]

Yield Strength (YS)

Yield strength or yield stress is the beginning of plastic deformation. The load required to permanently stretch a rod by 0.2 % of its original length is called yield strength.

A 100 cm rod, for example, that has been loaded so that it has a permanent stretch of 0.2 % has been permanently lengthened to 100.2 cm, when the load is removed.

Compressive Strength

Compressive stress in comparison to tensile strength is negative stress. Failure occurs as yield for ductile metals, whereas brittle materials, e.g. cast iron, will *shatter*. Fracture occurs at an oblique angle to the length of the sample. It is unlikely that a clean break will result; rather, several pieces will occur from compressing the material.

Stiffness

If the same tensile stress is applied to two materials, the stiffer of the two will lengthen less. Stiffness is defined by Young's Modulus (YM) or elastic modulus, the ratio of applied stress to the strain it produces in the material. The smaller the strain, the greater the stiffness.

Glossary

to shatter	to break suddenly into very small pieces

Task 2. Complete the table.

hard versus soft	equals	.... yield strength (resistance to plastic deformation) versus .... yield strength
ductile versus	equals	appreciable plastic deformation before fracture versus .... plastic deformation before fracture
stiff .... easily bent	equals	high .... versus low Young's Modulus

3.3 Important Properties for Manufacturing

One of the most important aspects in manufacturing is to choose the right material for a particular application. The properties, cost and availability of the material have to be considered.

When referring to metals in manufacturing, five properties are of importance:

- ductility
- durability
- elasticity
- hardness and
- *malleability*

Task 1. *Choose one of the above properties as an appropriate title for the paragraphs. Add the proper names to the chemical symbols.*

....

The metals are easy to form and stretch without breaking or fracturing and keep their new shape. Metals like Cu ...., Sn ...., Au .... and Ag .... all have this property and are often used to make, e.g. wire and tubing.

The same is true for soft low-carbon steels but high-carbon steels and cast iron soon fracture when stretched, as they are too brittle.

....

The metals can be stretched to some point, but go back to their original shape as soon as the stress is removed. Among metals, some steel alloys show this property, e.g. a high-carbon steel called spring steel. Other hard steels, e.g. tool steel and cast iron, can be stretched very little or not at all.

....

The metals can withstand friction. This characteristic makes them suitable for moving parts of machines and cutting edges of tools, e.g. steel alloys with a high C .... content.

....

These metals are easy to form without fracturing, and keep their new shape. Forming is done by, e.g. rolling or pressing, often with the application of heat. Au, Ag, Pb ...., Cu and low-carbon steel alloys belong to this group and are made into containers, wheels and, of course, jewelry.

Glossary

malleability	the property of sth that can be worked/hammered/shaped without breaking

Task 2. *Translate the following paragraph. You may need the terms in the box.*

> alloy; be in short supply; chromium; coat; coating; corrode; corrosion; durable, durability; paint; be resistant to; tungsten

Korrosionsbeständigkeit

Korrosionsbeständige Metalle korrodieren praktisch nicht, wenn sie Luft und Feuchtigkeit ausgesetzt sind. Cr und Pt verfügen über hohe Korrosionsbeständigkeit, sind aber teuer und knapp. Au, Ag und Al sind ebenfalls sehr korrosionsbeständig. As, Fe und Stahl korrodieren schneller und müssen deshalb mit einer Korrosionsschutzschicht versehen werden, z. B. durch einen Farbanstrich. Es gibt Stahllegierungen, die sehr korrosionsbeständig sind, z. B. Wolfram-Stahl, der aus W, Cr, C und Fe besteht.

3.4 Metal Alloys

A metal alloy is a metallic substance composed of two or more elements, which keep the same crystal structure in the alloy. Metals are combined with metals and/or with non-metal elements, for example carbon. Metal with metal alloys are made by mixing the molten substances and then cooling them until they solidify.

Common alloys are brass (copper + zinc) and aluminum alloys (aluminum + copper, aluminum + magnesium), and steel. Plain carbon steel contains only iron and carbon, while alloyed steels, e.g. stainless steel, contain chromium as the main alloying element.

Alloy systems are classified either according to the base metal, i.e. the metal serving as base of the alloy, or according to some specific characteristic that a group of alloys share.

Depending on their composition, metal alloys are often grouped into two classes:

      ferrous and non-ferrous alloys.

3.4 Metal Alloys

Ferrous Alloys

The principle constituent is iron as in, e.g. steel and cast iron. They are produced in larger quantities than any other metal type, being especially important as construction materials.

Iron and steel alloys can be produced using relatively economical techniques to be extracted, *refined*, alloyed and fabricated. Ferrous alloys have a wide range of physical and mechanical properties. However, they have relatively high density, which means they weigh a lot; their electrical conductivity is comparatively low and they are *susceptible* to corrosion in some common environments.

(from Callister, modified and abridged)

Glossary

ferrous	of or containing iron
to refine	to make/become free from impurities
to be susceptible to susceptibility, *n*	to be easily affected/influenced by

Nonferrous Alloys

Since nonferrous alloys have distinct limitations, other alloy systems are used for many applications, e.g. copper, aluminum, magnesium, titanium alloys, super alloys, the noble metals, and other alloys, including those that have nickel, lead, tin, zirconium and zinc as base metals.

(from Callister, modified and abridged)

Task 1. *Practice so-called chain questions. Ask a classmate a question about information provided by the texts above. The student who has answered the question asks another student a question, who answers and so on.*

Question: *What does the term metal alloy refer to?* Answer: *It refers to ...*

How ....?

....

Which ....?

....

What ....?

....

Why ....?

....

3.5 Case Study: Euro Coins

Figure 7: Euro coins

In deciding which metal alloys to use for the euro coins, their physical properties were an important issue.

Task 1. *Add captions to the following paragraphs.*

Required Characteristics

....

Differences in size and color help to distinguish *denominations* of coins which requires alloys to keep their distinctive color without *tarnishing*.

....

Coins should be difficult to *counterfeit*. Most vending machines use electrical conductivity to prevent false coins from being used. Thus, each coin has its own unique electronic signature, which depends on its alloy composition.

....

The alloy must be easily coined to allow design reliefs to be stamped into the coin surfaces.

....

Wear resistance against long-term use is necessary, to retain the reliefs.

3.5 Case Study: Euro Coins

....

In common environments it is required to ensure minimal material losses over the lifetimes of the coins.

....

Coins no longer fit for use should be recyclable.

....

The alloys should prevent undesirable microorganisms from growing on the coins' surface.

Selection of Alloys

As the base metal for all euro coins, copper was selected. Several different copper alloys and alloy combinations were selected for the different coins.

The 2 Euro Coin

A bimetallic coin, consisting of the silver-colored outer ring, a 75Cu-25Ni alloy, and the inner disk which is composed of a gold-colored, three-layer structure of high-purity nickel that is *clad* on both sides with a nickel brass alloy (75Cu-20Zn-5Ni).

The 1 Euro Coin

Also bimetallic; the alloys used for its outer ring and inner disk are reversed from those of the 2 euro coin.

The 50, 20 and 10 Euro Cent Pieces

These coins are made of so-called Nordic Gold alloy (89Cu-5Al-5Zn-1Sn).

The 5, 2, and 1 Euro Cent Pieces

These coins are made of copper-plated steel.

(from Callister, modified and abridged)

Glossary

denomination	a unit of value, esp. for money
to tarnish tarnish, *n*	to discolor a metal surface by oxidation, to become discolored
to counterfeit	to make a copy of sth, with criminal intent, to fake
to clad	to cover a material with a metal

3.6 Grammar: Adverbs I

Adverbs are frequently used in scientific writing, since they describe activities and characteristics. The way adverbs are formed and used in English differs considerably from other languages.

Task 1. *Complete the survey on adverbs and add examples.*

Formation of Adverbs

Add .... to an adjective.

slow – ....

Change adjectives ending in **-le** to ....

possible – ....

Change adjectives ending in **-y** to ....

sticky – ....

Change adjectives ending in **-ic** to ....

magnetic – ....

Irregular Forms

good – ....

hard – ....

(The form **hardly** exists, but it means ....)

fast – ....

friendly – ....

Use of Adverbs

Task 2. *Work in a group. Look through the texts about metals starting with 3.1. Make a list of the phrases that contain adverbs in combination with adjectives.*
Describe the use of adverbs in these phrases.

....

....

....

....

3.7 Case Study: The Titanic

Figure 8: The Titanic [wikipedia]

As is well known, the Titanic sank on her first trip across the Atlantic Ocean in 1912 after hitting an iceberg. 1,513 of the 2,224 people on board died, mainly because there were only 1,178 places in the ship's lifeboats. At the time of the collision, the Titanic was traveling at the relatively high speed of 22 knots, which equals 41 km/h, a dangerous speed at this time of the year, as icebergs are common in the North Atlantic in early spring. The *hull* of the Titanic was double-bottomed and divided into 16 compartments. As the ship would not sink even if four of these compartments filled with water, she was thought to be unsinkable.

After divers had found the wreck of the Titanic at a depth of about 13,000 ft (3,950 m) in 1985, a 1996 expedition used *sonar* imaging to discover a series of six narrow cuts in the hull. The damage totaled only 12 square ft, about the size of a human body, but the cuts were located 20 ft below the waterline, where water pressure forced the sea water through them at a rate of almost 7 t/s.

Researchers began questioning if poorly manufactured materials played a role in the ship's sinking. A major factor contributing to the disaster was the brittleness of the steel used.

Task 1. *Add the chemical symbols.*

Steel produced at the time the Titanic was built generally had a higher percentage of S (....................) and P (....................) than would be allowed today, resulting in steel that fractured easily. Samples of Titanic fragments were tested to determine the steel's chemical make-up, tensile strength, microstructure and grain size, as well as its responses to low temperatures. As the metallurgists had suspected, the steel was full of large MnS (....................) impurities that created weak areas and caused the metal to be brittle.

Under extreme conditions, such as the unusually cold, 28 F water temperatures of the North Atlantic at the time of the disaster, the steel became fragile and, subjected to the violent impact, immediately fractured.

Glossary

hull	the body of a ship
sonar	a system using transmitted and reflected underwater sound waves to detect/locate/examine submerged objects
t/s	tons per second

Task 2. *Read the text above, then decide whether the statements are true or false. Rewrite the false statements if necessary.*

Most passengers drowned because the ship sank fast.

....

Median speed for a cruise ship was 22 knots.

....

Divers found one deep cut in her hull.

....

Impurities in the steel were responsible for the poor performance of the Titanic's steel.

....

Glossary

median	relating to or constituting the middle value in a distribution, e.g. the median value of 17, 20 and 36 is 20

3.8 Grammar: The Passive Voice

The passive voice appears in scientific texts rather frequently. This is appropriate for an impersonal use of the language, where the acting person is of no importance and therefore does not have to be mentioned. The passive is also used to describe a process.

Formation of the Passive

The passive form of the verb consists of two parts:

the form of **be** in the appropriate form and tense

plus the past participle of the verb, i.e. the so-called third form, as in write –wrote – **written.**

3.8 Grammar: The Passive Voice

Task 1. *Fill in the missing verb forms*

Tenses of the Passive

Simple Present: simple present of be + past participle (p.p.) of the verb

    The article **is published** in *Nature*.

Present Progressive: simple present of be + being + p.p. of the verb

    The paper .... (print) right now, it can't be changed.

Simple Past: simple past of be + p.p. of the verb

    The book .... (edit) last month.

Present Perfect: present perfect of be + p.p. of the verb

    The article .... (publish) recently.

Past Perfect: past perfect of be + p.p. of the verb

    The draft .... (finish) before the lecture.

Future Tenses: future I or II of be + p.p. of the verb

    The hand-outs .... (copy) as soon as possible.

    The thesis .... (hand in) by now.

Conditional: conditional I or II of be + p.p. of the verb

    If universities received more money, more research .... (do)

    The report .... (write) by now, if the student had not gone skiing and broken his wrist.

3.9 Case Study: The Steel-Making Process

Figure 9: Steel-making machinery [wikipedia]

Task 1. *Work with a partner. Refer to* 3.8 Grammar: The Passive Voice. *Put in the verbs in brackets in the correct form.*

There is no single substance .... steel: there are dozens of different types of steel – of different compositions and with different properties. (call) "Ordinary" steel can .... as an alloy of iron containing a small but fixed amount (up to 1.5 %) of carbon. (describe) The many special steels which are available have several other metals .... in as well. (mix) The properties of steel depend not only on its composition but also on any heat treatment .... to it after manufacture. (give) *Pig iron*, with its high proportion of impurities, is too brittle for most purposes, and the bulk of what .... in *blast furnaces* .... into steel. (convert; produce)

The steelmaking process requires that, after most of the carbon and practically all of the other impurities (Si, S, P) .... by oxidizing, the right amount of each of the required elements .... (add; remove)

Of the main steelmaking processes .... today, the one by which most steel is manufactured is the basic oxygen process. (use) This method is fast and over 300 t of steel

3.9 Case Study: The Steel-Making Process

can .... in as little as 40 min. (produce) A converter, which is a huge steel, *pear-*............................ container, called vessel, of up to 300 t capacity, is mounted so that it can .... either way for charging and tapping. (move; shape) It is charged with .... pig iron from the blast furnace, along with up to about half of its mass of *scrap* iron or steel. (melt) A water-............................ tube, called lance, can .... vertically into the vessel, delivering a high powered jet of pure oxygen, thus burning the carbon .... in the iron. (cool; dissolve; lower) The impurities .... rapidly, (C to CO_2 and S to SO_2) and escape as gases. (oxidize)

Glossary

pig iron	crude iron
blast furnace	the oven in which ore is melted to gain metal
ore	a mineral from which a metal can be extracted
pear-shaped	having a round shape becoming gradually narrower at the end
to tap	to remove by using a device for controlling the flow of a liquid
scrap iron	metal objects that have been used

Chapter 4 Ceramics

4.1 Introduction

The term ceramic comes from the Greek word *keramikos,* which means burnt substance. The desirable properties of these materials are normally achieved through a high-temperature heat treatment called firing. Up until the past sixty years, the most important materials in this class were called traditional ceramics, for which the raw material is *clay*, e.g. *china*, *bricks*, tiles and in addition, glasses and high-temperature ceramics. Recently, significant progress has been made in understanding the fundamental character of these materials and of the *phenomena* that occur in them that are responsible for their unique properties. Consequently, a new generation of these materials has evolved, and the term ceramic has taken on a much broader meaning. These new materials are applied in, e.g. electronics, computers, communication technology, biomedical implants and aerospace.

(from Callister, modified and abridged)

Glossary

clay	a kind of earth that is soft when wet and hard when dry
china	high-quality porcelain, originally made in China
brick	a rectangular block of baked clay used for building
phenomenon, phenomena, *pl*	a fact/event that can be identified by the senses

Task 1. *Work with a partner. Translate the following sentences into German. Cover the German version and translate them into English. Compare the two English versions*

English	German	English
The Greek word *keramikos* shows that the desirable properties of these materials are normally achieved through a high-temperature heat treatment called firing.		
Traditional ceramics are those for which the primary raw material is clay.		

4.2 Structure of Ceramics

Ceramics are compounds between metallic and non-metallic elements. They are most frequently oxides, nitrides and carbides. A composite material of ceramic and metal is cermet. The most common cermets are cemented carbides, which are composed of an extremely hard ceramic, bonded together by a ductile metal such as cobalt or nickel. In addition, there are the traditional ceramics mentioned before, those composed of clay minerals, as well as cement and glass. As ceramics are composed of at least two and often more elements, their crystal structures are generally more complex than those of metals.

(from Callister, modified and abridged)

Task 1. *Read the text above and decide whether the statements are true or false. Rewrite the statements if necessary.*

Ceramics are non-metallic, inorganic materials.

....

Ceramics can be compounds of at least three elements.

....

....

4.3 Word Formation: Suffixes in Verbs, Nouns and Adjectives

The texts you have worked with so far contain nouns, adjectives and verbs with suffixes worth remembering. Most of them are of Latin origin and are typically used in scientific texts. Germanic suffixes, e.g. -en and -ship, appear as well.

Task 1. *Work in a group. Add examples with collocations, i.e. two or more words often used together. Scan previous or following texts to find collocations.*

suffix   example with collocation

-(a)tion	*plastic deformation*
-able/ -ible	
-al	
-ance -ence	
-ant -ent	
-ary	

-ate	
-en	
-ic/-ical	
-ify	
-ion -ition	
-ist	
-ity	
-ive	
-ize -ization	
-ment	
-ness	
-ous	
-ship	

Task 2. *Fill in the table, adding the appropriate preposition if necessary.*

noun	adjective	verb
arrangement	n.a (not applicable)	to arrange
atom		
		to apply for
		to bond
	n.a.	to configure
dependence		
example		
geometry		n.a.
		to interact with
		to notice
	soft	

noun	adjective	verb
	solid	
structure		
		to vary

4.4 Properties of Ceramics

Task 1. *Work with a partner. Fill the gaps in the text with words from the box in their correct form.*

> characteristic; conductivity; deformation; ductility; fracture; load; magnetic; strength

With regard to mechanical behavior, ceramic materials are relatively stiff and strong. Their stiffness and .... are comparable to those of the metals. In addition, ceramics are typically very hard. On the other hand, they are extremely brittle, i.e. lack ...., and are highly susceptible to fracture, which limits their applicability in comparison to metals. The principal drawback of ceramics is a *disposition* to catastrophic .... in a brittle manner with very little energy absorption. At room temperature, both crystalline and non-crystalline ceramics tend to fracture before plastic .... can occur in response to an applied tensile ....

Ceramics typically insulate against the passage of heat and electricity, i.e. they have low electrical ...., and they are more resistant to high temperatures and harsh environments than metals and polymers. With regard to optical ...., ceramics may be transparent, translucent or opaque, and some of the oxide ceramics, e.g. Fe_3O_4, exhibit .... behavior.

(from Callister, modified and abridged)

Glossary

| disposition | a physical property/tendency |

Task 2. *Define the following terms:*

transparent ....

translucent ....

opaque ....

Task 3. *Work with a partner. Match the German terms in the box with the corresponding English terms, and add statements about the properties of ceramics.*

| Anwendbarkeit; Anfälligkeit; Isolation |

Anwendbarkeit:

....

Anfälligkeit:

....

Isolation:

....

4.5 Case Study: Optical Fibers versus Copper Cables

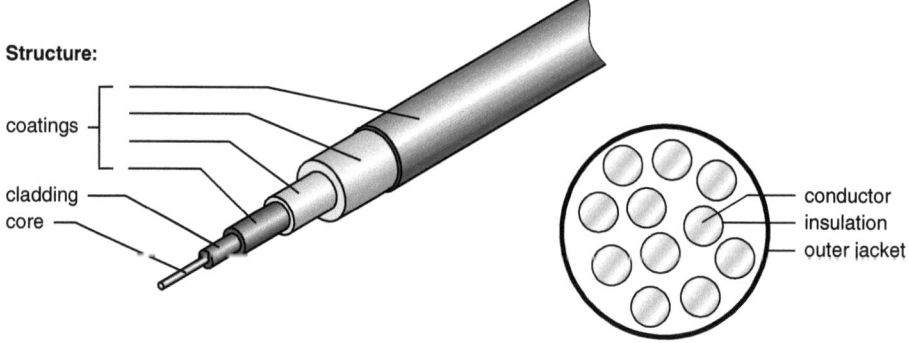

Figure 10: Optical fiber

Optical fibers, used in modern optical communication systems are an example for the application of an advanced ceramic material. They are made of extremely high-purity silica, which must be free of even extremely small levels of impurities and other defects that would absorb, scatter or weaken a light beam. Sophisticated processing has been developed to produce fibers that meet the rigorous restrictions required for this application, but such processing is costly.

4.5 Case Study: Optical Fibers versus Copper Cables

Optical fibers started to replace some uses of copper cables in the 1970s, e.g. in telecommunications and cable TV. In these applications they are the preferred material, because the fibers carry signals more efficiently than copper cable and with a much higher bandwidth, which means that they can carry more channels of information over longer distances. For optical fibers, the longer transmission distances require fewer expensive repeaters. Also, copper cable uses more electrical power to transport the signals. In addition, optical fiber cables are much lighter and thinner (about 120 micrometers in diameter) than copper cables with the same bandwidth so that they take up less space in underground cabling *ducts*. It is difficult to steal information from optical fibers and they resist electromagnetic interference, e.g. from radio signals or lightning. Optical fibers don't *ignite* so they can be used safely in *flammable* atmospheres, e.g. in petrochemical plants.

Due to their required properties, optical fibers are more expensive per meter than copper. In addition, they can't be *spliced* as easily as copper cable, thus special training is required to handle the expensive splicing and measurement equipment.

(from Callister, modified and amplified)

Glossary

duct	a pipe for electrical cables and wires
to ignite, ignition, n	to begin to burn, to cause to burn
flammable	easily ignited, capable of burning, inflammable
to splice, e.g. cables	to join two pieces at the end

Task 1. Work with a partner. Refer to 2.6 Grammar: Comparison. *Compare glass fibers to copper cables, listing the pros and cons of each material.*

4.6 Grammar: Adverbs II

In 3.6 Grammar: Adverbs I, the use of adverbs that modify the following adjective is introduced. Examples of such modifying adverbs appear in the texts about ceramics as well.

In addition, these texts contain examples of another use of adverbs, namely adverbs modifying a sentence.

Task 1. *Work in a group. Search the texts on ceramics to find examples of sentences with adverbs. Make a list of the phrases and name the modified element.*

Recently, significant progress has been made in understanding the fundamental character of these materials. (recently modifies the sentence).

....
....
....
....
....
....
....
....
....
....
....
....
....

4.7 Case Study: Pyrocerams

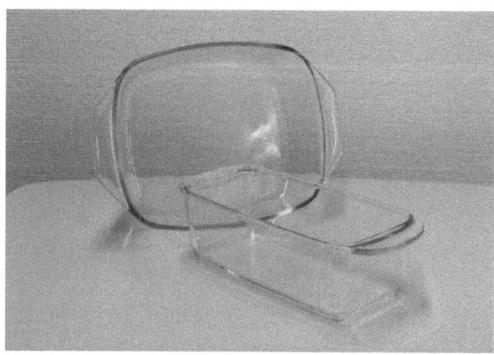

Figure 11: Ceramic cook ware

4.7 Case Study: Pyrocerams

Task 1. *Add captions to the following paragraphs.*

Pyrocerams or glass ceramics are widely used for ovenware, manufactured by, e.g. Corning-Ware or the German manufacturer Schott. The covalently bonded silicon carbide, silicon nitride and silicon aluminum oxynitrides, or sialons (alloys of Si_3N_4 and Al_2O_3), are the best materials for high-temperature structural use.

....

The *creep* resistance of the materials is outstanding up to 1300 °C, and their low thermal expansion and high conductivity make them resist thermal shock well in spite of their typically low toughness, the thermal shock resistance being better than that of most other ceramics. Pyrocerams exhibit excellent resistance to corrosion, which accounts for their use in the chemical industry.

....

These materials are manufactured by the high-temperature reaction of silicon nitride with aluminum oxide. They can be formed by hot pressing fine powders and sintering them in the process, or *slip casting* followed by pressureless sintering, which provides greater shape and manufacturing flexibility. If the constituents are varied, the properties of the final ceramic vary too. However, continuous exposure to high temperatures can result in the material's degrading back to these constituent parts.

....

Typical uses include burner and immersion heater tubes, injectors for nonferrous metals and protection tubes for nonferrous metal melting and welding fixtures.

(from Ashby/Jones, modified and amplified)

Glossary

creep, *n*	time-dependent permanent deformation of materials at high temperatures or stress
slip casting	the process of pouring liquefied material into a mold; after the liquid is drawn out, the solid is removed from the mold

Task 2. *Work with a partner. Reconstruct statements about high-temperature ceramics from the jumbled words without referring to the text. The first word is given.*

better ceramics is most of other resistance shock than that

Thermal ....

....

corrosion excellent exhibit resistance to too

Pyrocerams ....

....

and be by can fine formed hot powders pressing sintering them

They ....

....

are ceramics constituents final of properties the too varied vary

If ....

....

are best for high materials structural temperature the use

Sialons ....

....

ceramics for high include melting metal nonferrous of temperature tubes uses

Typical ....

....

4.8 Case Study: Spheres Transporting Vaccines

In order to find a way of delivering waterproof, time-release payloads of vaccines to the body, researchers at Cambridge Biostability Laboratory (CBL) in the UK studied the way body cells called osteoclasts remove *stray* bone fragments by attacking and dissolving them. Using calcium phosphate, the main mineral constituent of bone, the researchers developed *spheres* that can be slowly dissolved by osteoclasts, thus releasing the enclosed vaccine.

To build the spheres, a mixture of vaccine and calcium phosphate crystals in an *aqueous* solution is sprayed out of a *nozzle* into a stream of gas at around 170°C. The crystals are surrounded by a cloud of water molecules, which evaporate in the gas. As the water molecules evaporate, the crystals partially join together to form solid glassy spheres, five micrometer in diameter, with the vaccine embedded inside. The heat of the gas is absorbed by evaporative cooling before it can destroy the vaccine. The spheres prevent the vaccines from deteriorating or breaking down if not kept dry before release. They can be injected as a follow-up booster dose at the same time as the initial dose, releasing their contents over a period of months.

(from Biever, modified and abridged)

4.9 Useful Expressions for Shapes and Solids

Glossary

to stray	to move away from the place where sth/sb should be
sphere	a solid figure that is completely round
aqueous	watery
nozzle	a device with an opening for directing the flow of a liquid

Task 1. Read the text above then answer the following questions.

Why do researchers study the way the body removes bone fragments?

....

....

How are the embedded vaccines released from the spheres?

....

....

Why is the evaporation of the water molecules essential?

....

....

4.9 Useful Expressions for Shapes and Solids

Task 1. The table contains English terms for shapes. Add the corresponding adjectives and either draw the shape next to the term or write a short sentence that clarifies its meaning.

circle	
cone	
cube	
cylinder	
disc, *n.a.*	

ellipse
hemisphere
hexagon
pentagon
prism
rectangle
rhombus
semicircle, *n.a.*
sphere
square, *n*, *adj*
star-shape
trapezium
triangle

Chapter 5 Polymers

5.1 Introduction

Task 1. Work with a partner. Fill the gaps in the text with words from the box in their correct form.

> animal; application; cotton; industry; leather; molecule; plant; produce; property; rubber; silk; synthetic; wool

Naturally Occurring and Synthesized Polymers

Naturally occurring polymers, those derived from plants and animals, have been used for many centuries, for example wood, ...

Other natural polymers such as proteins, enzymes, *starches* and cellulose are important in biological and physiological processes in .... and .... With modern research tools it is possible to determine the molecular structures of these groups of materials and to develop numerous polymers that are *synthesized* from small organic .... referred to as *monomers*. .... polymers and, to a limited extent, biopolymers form the basis for plastics, rubbers, *thermosets*, fibers and adhesive and coating materials. Most monomers for such polymers are the products of the petrochemical .... For such applications, as well as for the structural function of some biopolymers in nature, adequate mechanical .... such as stiffness and strength are required. The synthetics can be .... inexpensively, and their properties may be controlled so that many are superior to their natural *counterparts*. In some ...., metal and wood parts have been replaced by plastics, which have satisfactory properties and may be produced at lower costs.

(from Callister, modified and abridged)

Glossary

starch	a white, tasteless powder found in plants, e.g. rice, potatoes
to synthesize, synthesis, *n*	to prepare a substance by chemical reaction

monomer	a molecule that can combine with others of the same kind to form a polymer
thermoset	a polymeric material that, once having cured or hardened by chemical reaction, will not soften or melt when heated
counterpart	here sth that has a similar function

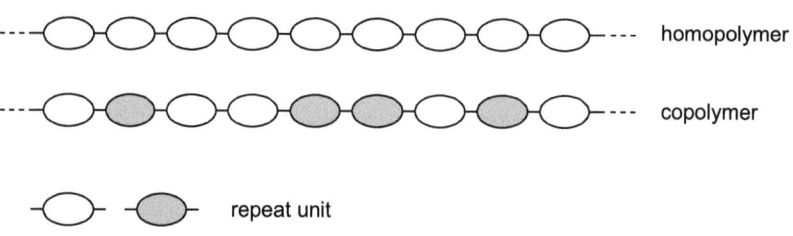

Figure 12: Structure of a homopolymer and a copolymer

Polymer can be defined as a substance whose molecules consist of many parts (Greek *poly* + *meros*). The term refers to molecules with many units joined to each other through covalent bonds, often repeating the units. That is why the units are called mers or repeat units. When the units are all of the same kind and joined together linearly, it is a homopolymer, whereas a copolymer has more than one type of repeat unit. Polymers can contain up to several hundreds or thousands of repeat units. Because of the resulting long chain, high molecular weight and large size, these polymers are called macromolecules. Polymers can be named on the basis of the monomer(s) from which they are derived by adding the prefix poly- to the monomer. Alternatively, a polymer can be named on the basis of its repeat unit structure. Complex biopolymers, e.g. cellulose, or synthetic polymers are often referred to by their trivial name, e.g. Nylon 6,6, the structure-based name of which is poly(hexamethylene adipamide).

Task 2. *Work with a partner. Draw a diagram of the chain structure of polyethylene with its repeat units.*

5.2 Word Formation: The Suffix -able/-ible

Adjectives ending in **-able/-ible** are often used in scientific texts, as they can replace longer verbal phrases:

> The specimen exhibits elongation **that can be appreciated**.
>
> The specimen exhibits appreci**able** elongation.

The suffix –able is derived from 'to be able to do sth' and can mean that something can be done. The form -able also occurs in the form -ible as in non-reversible, meaning 'cannot be reversed'. As the two forms are pronounced in almost the same way, they are often confused in spelling.

Task 1. *Work in a group. Form adjectives with the suffix -able/-ible that belong to the same word family as the verbs in the box. Add a suitable noun to form a collocation.*

access; appreciate; attribute; compare; desire; flex; notice; perceive; rely; reproduce; suit

-able	-ible
	access: make science *accessible to* all students

5.3 Properties of Polymers

Task 1. *Add the names of the polymers.*

Some of the common and familiar polymers are PE (..), Nylon, PVC (..), PC (..), PS (..) and silicone rubber. Polymers typically have low densities. Except for so-called high-performance polymers they are not as stiff or as strong as ceramics or metals. However, considering the polymers' low densities in comparison to metals and ceramics, their stiffness and strength on a per mass basis are equal or even superior to metals and ceramics. Many polymers are extremely ductile and pliable, thus they are easily formed into complex shapes. In general, they are relatively inert chemically (do not react with other substances) and are unreactive in a large number of environments. One major drawback to polymers is their comparatively poor heat stability. The tendency to soften and/or *decompose* at modest temperatures in some instances limits their use. Furthermore, they have low electrical conductivities and are nonmagnetic, features which may prove to be of advantage.

(from Callister, modified and abridged)

Glossary

to decompose	to change chemically, to decay

Task 2. *Make a list of the properties of polymers as mentioned in the text. Then name a property and ask a student in the class to give an explanation and/or additional information.*

Student 1 states: "Polymers show poor heat stability."

Student 2 adds: "This means they tend to …"

....
....
....
....
....
....
....
....
....
....

5.4 Case Study: Common Objects Made of Polymers

Figure 13:
Objects made of polymers

Task 1. *Work with a partner. Describe the required material properties of four common objects: billiard balls, bike helmets, plastic spoons, water bottles.*

5.5 Case Study: Ubiquitous Plastics

Plastics today

Uta Scholten, of the German Plastics Museum Association in Düsseldorf says: "Most people today don't know there was a time before plastics." This was a time when a soccer ball still was made of leather, not foamed PU, and a surfboard was made of wood not PE.

Today, yogurt tubes are made of PS, CDs of PC, shoes of PU, waste baskets of PP, computer keyboards of ABS (a copolymer of acrylonitrile, butadiene and styrene), and soda bottles of PET poly(ethylene terephthalate). These materials, called plastics in English, were given the name Kunststoffe by the German chemist Dr. Ernst Richard Escales in 1910, later also referred to as Plastik in a critical way. But over the last few years they have shaken off their image as cheap or inferior substitutes. "These days, plastics have a high-quality image," says Dirk Ziems, manager of a market research institute in Köln, Germany. "The elegant appearance of the iPod cannot be topped, and the functionality of modern athletic clothing will not be surpassed soon."

Plastics in architecture, fashion and design

The Swiss architects Jacques Herzog and Pierre de Meuron gave the Allianz Arena in Munich an inflatable covering made of EFTE (ethylene – tetrafluoroethylene copolymer) plastic that can be illuminated in white, blue and red, the colors of Munich's two professional soccer teams.

The Allianz Arena consists of 66,500 square meters of EFTE film, 0.2 mm thick, cut into rhombus-shaped cushions. Fans inflate the cushions, which have an estimated service life of 25 years. Karsten Moritz from Rosenheim who engineered the arena's plastic façade is convinced that film skins give architects new opportunities, especially when combined with sophisticated technologies, such as liquid crystal layers that can be laminated with film, or the special effects created when light hits the edges of the film.

Fashion is another field with its sight set on plastics. Fashion guru Karl Lagerfeld surprised an interviewer by naming not *velvet* or silk as his favorite material, but plastics.

According to the local newspaper of San Francisco, the Chronicle, "Plastic furniture has become the focal point in some of the most elegantly designed rooms." The Prada Store in Beverly Hills, designed by Rem Kohlhaas, has wall coverings made of spongy, translucent PU mats. Spaces for items on display are simply cut out as needed. "No other material can be so lightweight and luminescent," says the designer.

Plastics in aircraft engineering

Jets have to be safe and airlines need planes that can fly economically. Consequently, the percentage of plastics integrated in jet planes is rising steadily. The development of the giant Airbus 380 has taken the use of plastics to a new level. For the first time in civil aviation, fiber composites were used to build wing boxes, which are the heart of any jet. Compared to a conventional aluminum structure, fiber composites help to reduce the total weight by 1.5 tons, which reduces fuel consumption while increasing payload and range. In comparison with the new jumbo jet, the proportion of plastics in an older Boeing is less than 5 % of the total weight. The A380 brings the figure up to 20 %, and in the Boeing 787, plastics make up more than half of the material used.

Plastics as a Commodity

For commodity manufacturers, plastic has become the material of choice for getting ahead of the competition. With its brightly colored iMac models, Apple proved that computers don't have to be gray boxes. However, the greater the demands imposed by industry on plastics, the more expensive their manufacturing becomes. For this reason, industry is called on to develop corresponding methods that make the cost of manufacturing equal to or less than that of metallic materials.

(from Bayer *MaterialScience*, modified and abridged)

Glossary

| velvet | a type of cloth with a thick, soft surface |

Task 1. *Work with a partner. Match the following terms with the definitions.*

commodity ....

cushion ....

foam ....

luminescent ....

payload ....

spongy ....

ubiquitous ....

definitions:

bubbles of air together in a mass
emitting light
found everywhere
merchandise
resembling an artificial or natural material that is soft, light and full of holes
soft, protective pad
total weight an airplane can carry

Task 2. *Work with a partner. Make a list of plastic objects and their characteristics mentioned in the text. Refer to architectural design, interior design and aircraft engineering.*

5.6 Grammar: Reported Speech (Indirect Speech)

When reporting what another person said, the so-called back shift of tenses is often used.

If the reporting verb, e.g. to say, add, state, answer, is in the **past**, the verb in the reported clause in most cases shifts back into a form of the past.

Direct Speech:

Uta Scholten said: "Most young visitors of the museum **do** not know much about plastics."

Indirect Speech:

Uta Scholten **said** that most young visitors **did** not know much about plastics.

Formation and Use of the Back Shift

Task 1. *Back shift the verb in the reported sentence.*

Back shift of simple present to simple past

He said: "I **know** this author well."

He mentioned that he .... this author well.

Back shift of simple past and present perfect to past perfect

She said: "The first time I **read** about recycling plastics **was** forty years ago.

She stated that the first time she .... about recycling plastics .... forty years ago.

She added: "But I **have been interested** in recycling all my life."

She added that she .... in recycling since then.

Back shift of will to would

He said: "I **will** know more about the experiment next week."

He mentioned he .... more about the experiment the following week.

No Back Shift is Used

For statements of universal truths or irreversible facts.

He stated that the earth **turns** around the sun.

Task 2. *Work with a partner. Change some of the quotations in 5.5 from direct to reported speech and use different reporting verbs or expression.*

Dirk Ziem

....

....

....

San Francisco Chronicle

....

....

....

Rem Kohlhaas

....

....

....

5.7 Polymer Processing

Plastics can be shaped in many ways, e.g. some polymeric materials can be cast like metals, i.e. a molten material is poured into a mold and allowed to solidify. This process can be applied for both *thermoplastic* and thermosetting plastics, the latter being then *cured* in the mold to become the thermoset.

Glossary

thermoplastic, *n, adj*	a polymer that softens when heated and hardens when cooled
to cure	to improve the properties of polymers and rubber by combining with, e.g. sulfur under heat and pressure; cf. to vulcanize

Extrusion

Thermoplasts can also be extruded. Plastic chips are filled in a chamber containing a screw. The polymer is then heated by heating elements so that it melts. The screw forces the resulting resin through a *die*, which forms it into a special shape and lets the material cool.

This kind of processing produces, e.g. *tubes*, pipes, *rods*, and sheets or films.

Glossary

die	here: a metal block containing small holes through which the polymer is forced
tube	a long hollow pipe through which liquids/gases move
rod	a thin, straight piece or bar

Task 1. *Work with a partner. Read the text above. Then draw a schematic diagram of an extruder.*

Task 2. Work with a partner. Fill the gaps in the text with words from the box in their correct form.

article; eject; manufacture; metal ; pressure; shape; solidify

Injection Molding

Injection molding is used to .... both, thermoplastic and thermosetting materials. The first steps are the same as in extrusion. The molten polymer is injected at high .... into the mold and kept under pressure, until it has .... Then the mold is opened and the piece .... The molds are made from ...., usually either steel or aluminum, and .... to the desired form of the finished ...., e.g. garden chairs.

Task 3. Use the verbs in the box and the notes to write a text about blow molding.

blow in; cool; eject; extrude; fit; melt; place; produce; shape; use

Blow Molding:
plastic containers and bottles
hollow tube
in semi-molten state into cooled metal mold
air or steam under pressure
tube walls to contours of mold
hollow bottle or container

5.8 Case Study: Different Containers for Carbonated Beverages

Figure 14:
Carbonated beverage containers

Task 1. *Work in a group. Scan the text, then discuss and decide which material you would choose as manufacturer and as consumer for containers for carbonated beverages. Give reasons.*

A common item that represents some interesting material property requirements is a container for carbonated beverages.

The Material of Choice

should provide a barrier to the passage of carbon dioxide (CO_2), which is under pressure in the container;

must be nontoxic, unreactive with the beverage (including carbonic acid from dissolved CO_2), and preferably be recyclable;

should be relatively strong and capable of surviving a drop from a height of several feet when containing the beverage;

should be inexpensive, and the cost to fabricate the final shape should be relatively low;

should keep its optical clarity if optically transparent;

should be capable of being produced having different colors and/or labels

All three of the basic material types, metal (aluminum), ceramic (glass), polymer (PET) are used. They are all non-toxic and unreactive with the contained beverages. In addition, each material has its pros and cons.

Aluminum alloy is relatively strong but easily damaged. It is a very good barrier to the *diffusion* of CO_2 and can easily be recycled. The beverages are cooled rapidly and labels may be painted onto its surface. On the other hand, the cans are optically opaque and relatively expensive to produce.

Glass is a very good barrier to the diffusion of CO_2 and a relatively inexpensive material. It may be recycled, but it cracks and fractures easily and glass bottles are relatively heavy.

Plastic is relatively strong and can be made optically transparent. It is inexpensive, lightweight and recyclable. But plastic is not as good a barrier to the diffusion of CO_2 as aluminum and glass.

(from Callister, modified and abridged)

Glossary

diffusion	the movement of atoms/molecules from an area of higher concentration to an area of lower concentration

Your choice material as manufacturer:

....

....

....

....

....

....

....

Your choice material as consumer:

....

....

....

....

....

....

Chapter 6 Composites

6.1 Introduction

Task 1. *Work with a partner. Fill the gaps in the text with words from the box in their correct form.*

artificial; aerospace; bone; cellulose; corrosion; dissimilar; phase; transportation; underwater; wood

A number of composites occur in nature: .... consists of strong and flexible .... fibers surrounded and held together by a stiffer material called lignin. .... is a composite of the strong yet soft protein collagen and the hard, brittle mineral apatite. Yet many modern technologies require materials with unusual combinations of properties that cannot be met by natural composites or the conventional metal alloys, ceramics and polymeric materials. This is especially true for materials that are needed for ...., ... and .... applications. Aircraft engineers for example, are increasingly searching for structural materials that have low densities, are strong, stiff and resistant to *abrasion* and *impact* as well as ...., a rather impressive combination of characteristics. The problem is that strong materials frequently are relatively dense, i.e. heavy. Increasing the strength or stiffness typically results in a decrease in impact strength.

Generally speaking, a composite is considered to be any .... made multiphase material that shows properties of both *constituent phases* so that a better combination of properties is realized. The constituent phases in a composite are chemically .... and separated by a distinct interface. Many composite materials are composed of just two ...., the one phase being the matrix, which is continuous and surrounds the other phase, which is often called the *dispersed* phase.

The properties of composites are a function of the properties of the constituent phases, their relative amounts and the geometry of the dispersed phase, which means the shape, particular size, distribution and orientation of the particles.

(from Callister, modified and abridged)

Glossary

abrasion, to abrade	the process of being rubbed away by friction, to rub away
abrasive, *n, adj*	a substance that abrades, abrading
impact	a high force or load acting over a short time only
constituent phase	one of the *phases* from which a substance is formed
phase	a form or state of matter (solid/liquid/gas/plasma) depending on temperature and pressure
interface	the area between systems where they come into contact with each other
to disperse, dispersion, *n*	to distribute particles evenly through a medium

Task 2. *Work with a partner. Answer the following questions.*

What is the number of individual materials a composite is composed of?

....

What is the design goal of a composite?

....

....

6.2 Case Study: Snow Ski

A modern ski is a relatively complex composite structure, consisting of many parts, being composed of different materials:

the base: compressed carbon (carbon particles embedded in a plastic matrix); hard and abrasion resistant; provides appropriate surface

the top: ABS plastic having a comparatively low *glass transition* temperature; used for controlling and cosmetic purposes

the core: polyurethane plastic; acts as a filler

the core wrap: bidirectional layer of fibreglass; functions as a *torsion* box and bonds outer layers to core

the side: ABS plastic, cf. top

the edge: hardened steel; facilitates turning by cutting into snow

the *damping* layer: polyurethane; improves *shatter* resistance

(from Callister, modified and abridged)

6.2 Case Study: Snow Ski

Glossary

glass transition temperature T_g	the temperature at which, upon cooling, a non-crystalline ceramic or polymer transforms from a *supercooled* liquid to a solid glass
supercooled	cooled to below a phase transition temperature without transforming
torsion, torsional, *adj*	the stress/deformation caused when one end of an object is twisted in one direction and the other end is twisted in the opposite direction
to damp(en)	to make sth less strong, to soften
to shatter	to suddenly break into pieces

Task 1. Work with a partner. Draw the cross-section of a snow ski, showing the different layers of the composite structure as described.

Task 2. *The notes on a snow ski contain several expressions that can be used to describe a purpose. Make a list of the expressions. Then use them in sentences.*

....
....
....
....
....
....
....
....

6.3 Grammar: Gerund (-ing Form)

The gerund (after Latin *gerundium*), also called **–ing** form, is identical in form to the present participle as in the sentence:

Talk**ing** to Mr. Brown, she left the room.

In this sentence, the present participle *talking* stands for *while she was talking*, and is used to abbreviate the sentence. Some linguists do not differentiate between the gerund and the present participle, but most English grammar books explain the usage of the gerund in a separate chapter.

Formation of the Gerund

Task 1. *Fill in the missing forms.*

Add -ing to the infinitive of a verb.

to avoid – ....

Drop the end-e.

to freeze – ....

Double the final consonant when it is preceded by a stressed vowel.

to stop – ....

6.3 Grammar: Gerund (-ing Form)

Use of the Gerund

The gerund can be used like a noun, and it can be modified by determiners like direct and indirect articles (the, a), or pronouns (my, your).

The freezing of water is one of the most common transformations in nature.

The gerund can be used like a verb and have an object.

They finally stopped questioning all information.

Note that some verbs can be used with both the gerund and the infinitive with a change in meaning. These verbs and examples are listed in any English grammar book.

Gerund after Prepositions

Task 2. *Work with a partner. Use the gerund and form meaningful sentences with the prepositions from the box and the following phrases.*

after; before; by; of; on; to; without

alter the size of the sample – increase the temperature

After/before altering the size of the sample, the temperature was increased.

in spite - study hard – not pass the exam

....

look forward – finish the academic year

....

the edge of the ski – facilitate turning - cut into the snow

....

see the new instrument - enter the lab

....

start the instrument – read the manual first

....

Gerund after Adjectives + Preposition

Task 3. *Add the prepositions from the box, some of which will have to be used several times, and change the verbs into the gerund.*

about; against; at; for; in; of; on; to; with;

She is good/bad .... (work) with students.

He is angry .... (lose) his notebook,

Professor X. is disappointed .... (see) such a bad report.

The instruments are famous .... (give) reliable performance.

The company is interested .... (hire) him.

Gerund after Nouns + Preposition

This is the advantage .... (use) underground cable.

Special clothing protects against the danger .... (be) exposed to radiation.

Gerund after Verbs + Preposition

He was accused .... (plagiarize) from the internet.

The research group concentrates .... (develop) applications for new composites.

Students have to cope .... (solve) many problems.

Medical interns have to get used .... (work) long hours.

They decided .... (use) non-recyclable materials.

Gerund after Certain Verbs

Note that certain English verbs require a following gerund.

Lists of such verbs are listed in any English grammar book. Below are a few examples.

Task 4. *Use the gerund of the verbs in brackets to form meaningful sentences with the verbs from the box and the following phrases.*

admit; avoid; consider; include; justify; suggest

The task .... (write) an essay.

He had to .... (pay) that much for the chemical.

Please try to .... (expose) the sample to light.

We .... (vary) the temperature and frequency.

She .... (have, miss) this aspect of the material's failure.

The manual .... (work) under the exclusion of oxygen.

6.4 Case Study: Carbon Fiber Reinforced Polymer (CFRP)

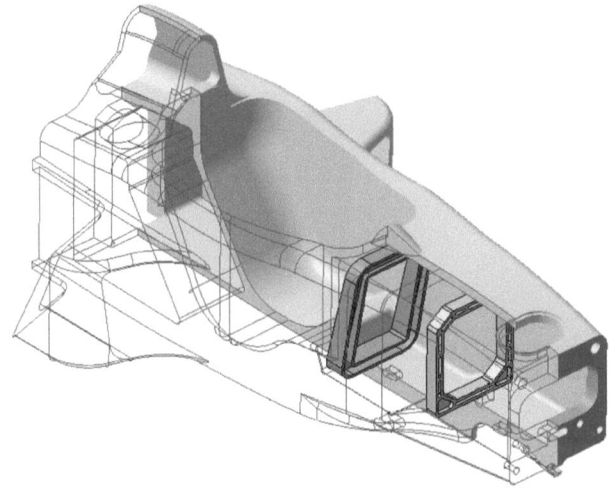

Figure 15:
Cross-section of the safety cell of a race car
[M. Trzesniowski]

This composite material is commonly referred to by the name of its reinforcing fibers, namely carbon fibers. To manufacture, e.g. body parts for race cars, carbon fibers are embedded as reinforcement into a matrix, which usually is epoxy. This is done by layering sheets of carbon fibers into a mold in the shape of the final product, the arrangement of the cloth fibers depending on the desired strength and stiffness properties of the product. The mold is then filled with epoxy and heated or air cured.

CFRP is a technologically important material. It is very strong and light-weight, non-corroding, heat-resistant, will not *ignite* and shrinks very little when exposed to high temperatures. Unfortunately, carbon fibers are expensive to manufacture.

Glossary

to ignite	to start to burn, make sth start to burn

Task 1. *Work with a partner. Read the text above. Then answer the question in a few sentences. Add anything you know about the subject.*

Why is CFRP used in racecar and, to some extent, mainstream car manufacturing?

....
....
....
....
....
....
....
....
....
....

6.5 Word Formation: Prefixes

The texts you have worked with so far contain prefixes worth noticing,

e.g. in nouns (**sur**face), adjectives (**in**combustible) and verbs (to **com**press).

Most prefixes are of Latin origin, which is typical of a scientific text, but there are also Germanic prefixes, e.g. to **em**bed.

Task 1. *Work in a group. Match the words from the box to the prefixes in the table. Add a collocation as well.*

| activate; atomic; author; band; calculate; change; clinic; coloured; compatible; conductor; cooled; crystallization; directional; due; electric; estimated; ethylene; fabricated; ferromagnetism; formation; friendly; function; functional; gram; immune; light; linear; measure; meter; metrical; molecular; notice; purity; similar; size; space; standing; structured; tube; type; typical; watt; zero |

6.5 Word Formation: Prefixes

prefix	collocation
a-	*atypical behavior*
aero-	
anti-	
auto-	
bi-	
bio-	
co-	
counter-	
de-	
di-	
dis-	
eco-	
ex-	
geo-	
im-	
inter-	
kilo-	
macro-	
mal-	
mega-	
micro-	
milli-	
mis-	
multi-	
nano-	
non-	
out-	

over-	
poly-	
pre-	
proto-	
re-	
semi-	
sub-	
super-	
trans-	
tri-	
ultra-	
under-	
uni-	

Chapter 7 Advanced Materials

7.1 Introduction

Task 1. *Work with a partner. Write an outline of the following presentation about advanced materials. Then give a short presentation on the basis of this outline. Take turns.*

"Good afternoon, Ladies and Gentlemen,

The topic of my short presentation today will be an introduction to advanced materials.

First, I am going to discuss two material types that belong to this category. Second, I will mention current applications of advanced materials.

Advanced materials can be of all material types, e.g. metals, ceramics and polymers.

To obtain advanced materials, properties of traditional materials have been improved, that is significantly changed in a controlled manner. Advanced materials include semiconductors, biomaterials as well as smart materials and nano-engineered materials.

Two important classes of advanced materials I want to introduce here are smart materials and nano-engineered materials. Smart materials respond to external stimuli, such as stress, temperature, electric or magnetic fields. By way of example, consider shape memory alloys or shape memory polymers, which are thermo responsive materials, where deformation can be induced and recovered through temperature changes, as can be seen in this figure.

As I have already mentioned, advanced materials also include nano-engineered materials which have unique properties. These properties arise from structural features which are of nanoscale dimensions, i.e. 1 to 100 nanometers. A prominent example are carbon nano-tube filled polymers which can be employed as electrically conducting materials or high performance materials. Please refer to the next diagram showing room temperature electrical conductivity ranges of these polymers.

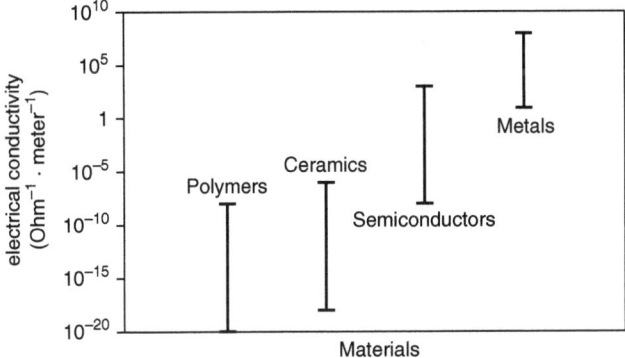

Figure 16: Room temperature electrical conductivity ranges for metals, ceramics, polymers and semi-conducting materials

Having looked at two classes of smart materials, I will now turn to some applications. Advanced materials are used in high-tech applications for, among others, lasers, *integrated circuits*, magnetic information storage, and liquid crystal displays (LCDs). They function in everyday electronic equipment such as computers, camcorders, or CD/DVD players. But advanced materials also operate in state-of-the-art devices for spacecraft, aircraft, and military *rocketry*.

In conclusion we have seen the structural versatility and wide range of potential applications of advanced materials. This is why they are being investigated in academic and industrial research laboratories world wide, and further developed and optimized for various tasks in industry.

Thank you for your attention, Ladies and Gentlemen. I'll be pleased to answer questions now."

(data from Callister, modified and abridged)

Glossary

integrated circuit	millions of electronic circuit elements incorporated on a very small silicon chip
rocketry	the science and technology of rocket design, construction and flight

7.2 Semiconductors

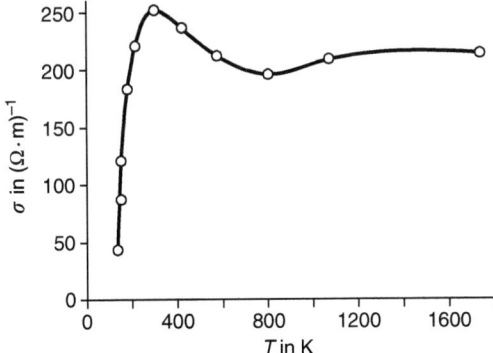

Figure 17:
Temperature dependence of electrical conductivity for semiconductors

Task 1. *Fill in the names of the elements.*

Semiconductors may be either elements, namely Si (..........................) and Ge (..........................), or covalently bonded compounds. Si is used to create most semiconductors commercially.

A semiconductor is a solid material with electrical properties that are intermediate between the electrical conductors such as metals and metal alloys and insulators, namely ceramics and polymers. The electrical characteristics of these materials are extremely sensitive to temperature and *minute* concentrations of *impurity* atoms, called doping. Depending on the type of the impurity, the impurity atom either adds an electron or creates a hole, i.e. a site where one electron is missing.

Intrinsic Semiconductors

The electrical properties are inherent in the pure material, and electron and hole carrier concentration are equal. With rising temperatures, the intrinsic electron and hole concentration increases dramatically.

Extrinsic Semiconductors

An extrinsic semiconductor has been doped, giving it different electrical properties from the intrinsic one. The electron and hole carrier concentration at thermal equilibrium has been changed. For extrinsic semiconductors, with increasing impurity dopent content, the room temperature carrier concentration increases whereas carrier mobility diminishes.

(from Callister, modified and abridged)

Glossary

minute	extremely small
impurity atoms	here atoms of a substance that are present in a different substance

Task 2. *Work with a partner. Write questions that elicit the answers contained in the sentences. Different questions are possible. Practice questions and answers with a partner, then switch roles.*

Which element is most often used to create semiconductors commercially?	Si is used to create most semiconductors commercially.
	Semiconductors have electrical properties that are intermediate between electrical conductors and insulators.
	The electrical characteristics of these materials are extremely sensitive to the presence of impurity atoms.
	The intrinsic electron and hole concentration increases dramatically with rising temperatures.
	Semiconductors are classified as either intrinsic or extrinsic on the basis of their electrical behavior

7.3 Case Study: Integrated Circuits

Task 1. *Work with a partner. Fill the gaps in the text with words from the box in their correct form.*

advancement; approach; consume; electronic; improvement; manufacture; miniaturize; perform

In electronics, an integrated circuit, also known as IC or microchip, is a .... electronic circuit consisting mainly of semiconductor devices as well as passive components. These circuits are .... on the surface of a thin substrate of semiconductor material. ICs revolutionized the world of electronics and nowadays appear in almost all .... equipment. Integrated circuits were made possible by discoveries which showed that semiconductor devices could .... the functions of *vacuum tubes*. Thanks to technological .... in semiconductor device fabrication in the mid 20th century, large numbers of tiny transistors could be integrated into a small chip.

This was an enormous .... over the *manual assembly* of circuits. The fact that reliable integrated circuits could be mass produced using a building-block .... in circuit design resulted in the fast adoption of standardized ICs in place of designs using transistors. The cost of integrated circuits is low because of mass production and because much less material is used. Being small and close together, the components switch quickly and .... less power than their discrete counterparts. In 2006, chip areas ranged from a few square millimeters to around 350 mm^2, with up to 1 million transistors per mm^2.

Glossary

vacuum tube	an electron tube from which all or most of the gas has been removed, letting electrons move without interacting with remaining gas molecules
manual assembly	putting together manufactured parts to make a completed product by hand

7.4 Grammar: Subordinate Clauses

Subordinate clauses are phrases that give answers to questions like Why? What … for?
    Why are impurity atoms added to these materials?
    Impurity atoms are added **in order to influence electrical properties.**

Expressions Introducing Subordinate Clauses
in order to/so as to + the infinitive of the verb
    The properties of the material were changed in order to/so as to improve performance.
so that
    The properties of the material were changed so that performance improved.
for + noun + to + infinitive
    For the metal to melt, higher temperatures must be used.

Task 1. Rewrite the following sentences, using the expressions in brackets.

Scientists planned to make possible the development of integrated circuitry. That's why they introduced semiconductors. (in order to)

....

....

The audience stayed in the lecture hall because they wanted to be able to hear the second lecture. (so that)

....

....

Researchers added impurities, because conductivity had to be optimized. (so as to)

....

....

Circuit breakers were installed, because one did not want the system to overload. (for ... to ...)

....

....

7.5 Smart Materials

Task 1. Work with a partner. Translate the following text into English.

Intelligente Werkstoffe sind in der Lage, Veränderungen in ihrer Umgebung zu erkennen und auf derartige äußere Impulse auf festgelegte Weise zu reagieren. Ähnliche Eigenschaften finden sich bei lebenden Organismen.

Intelligente Werkstoffe haben einen Sensor, der ein Eingangssignal erkennt, und einen Aktuator, der eine entsprechende Reaktion und Adaptation auslöst.

Der Aktuator kann als Reaktion auf eine Veränderung von Temperatur, Druck, Licht, oder eines elektrischen bzw. magnetischen Felds eine Veränderung z. B. der Form, Position, oder mechanischer Eigenschaften hervorrufen.

Smart Materials ....

....

....

....

....

....

....

....

7.5 Smart Materials

Task 2. Work with a partner. Reconstruct the text about materials for actuators from the jumbled sentence parts in the brackets.

Materials Used for Actuators

Shape Memory Alloys

Shape memory alloys ... (alloys can consist metal of or polymers)

Shape memory alloys can consist of metal alloys or polymers.

These alloys are thermo-responsive materials, where deformation can be ... (caused changes deformation temperature through).

....

....

After having been deformed, they return to ... (changed is original shapes temperature the their when).

....

....

Piezoelectric Ceramics

Piezoelectric ceramics expand and contract in response to an applied electric field or voltage; they also generate ... (altered an are dimensions electric field their when)

....

....

Magnetostrictive Materials

The behavior of magnetostrictive materials is analogous to that of the piezoelectrics, except that ... (fields magnetic respond they to)

....

....

Electrorheological/Magnetorheological Fluids

Electrorheological/magnetorheological fluids are two types of fluids whose properties, e.g. viscosity, can be changed ... (an applying by electric field magnetic or)

....

....

(from Callister, modified and abridged)

7.6 Nanotechnology

The history of science shows that, to understand the chemistry and physics of materials, researchers generally have begun by studying large and complex structures and then later investigated smaller fundamental building blocks of these structures.

However, *scanning probe microscopes*, which permit observation of individual atoms and molecules, make it possible to manipulate and move atoms and molecules to form new structures and thus design new materials that are built from simple atomic-level constituents, an approach called 'materials by design'. This ability to arrange atoms provides opportunities not otherwise possible to develop and study mechanical, electrical, magnetic and other properties. In the term nanotechnology, the prefix nano denotes that the dimensions of these structural entities are on the order of a nanometer (10^{-9} m). As a rule, they are less than 100 nanometers (equivalent to approximately 500 atom diameters).

(from Callister, modified and abridged)

Glossary

scanning probe microscope	(SPM), a microscope that scans across the specimen surface line by line, from which a topographical map of the specimen surface (on a nanometer scale) is produced

Task 1. *The text refers to two kinds of scientific approaches, the top-down and the bottom-up approach. Explain.*

In the so-called top-down approach to the chemistry and physics of materials, researchers study ....

....

....

In the so-called bottom-up approach, ...

....

....

7.7 Case Study: Carbon Nanotubes

Task 1. *Work with a partner. Fill the gaps in the text with words from the box in their correct form.*

applicable; atom; consist; diameter; ductile; efficient; end; field; know; molecule; thickness

7.7 Case Study: Carbon Nanotubes

The structure of a nanotube .... of a single sheet of graphite, one atom in ...., which is rolled into a tube. At least one .... of the tube is capped with a C_{60} *fullerene* hemisphere. Each nanotube is a single .... composed of millions of .... The length of the molecule is thousands of times greater than its .... Nanotubes are extremely strong and stiff and relatively .... For single-walled nanotubes, tensile strengths range between 50 and 200 GPa, which is the strongest .... material so far. Nanotubes have unique electrical properties and are .... conductors of heat. Because of their unique properties, nanotubes are extremely useful as reinforcement in composite materials and will be .... in many ways in nanotechnology, electronics, optics and other .... of materials science.

(from Callister, modified and abridged)

Figure 18: Carbon nanotube structure

Glossary

fullerene	carbon molecule named after R. Buckminster Fuller, sometimes called buckyball, composed entirely of C in the form of a hollow sphere, ellipsoid or tube

7.8 Grammar: Modal Auxiliaries

Scientific texts use constructions with modal auxiliaries, also called 'modals', e.g. when the texts are about a potential future development or when hypothetical statements are made.

Formation and Use of Modal Auxiliaries

Modals require the verb in the infinitive.

    Solar energy **could** significantly **reduce** consumption of oil in coming decades.

Modals do not add do/does/did in questions or in negative sentences.

    Fuel cells **may not** provide enough energy to sufficiently reduce fuel consumption.

Modals have no past or future form (except for could and would).

Modals and their Meanings

can and **could** express

the ability and the permission to do sth, cf. to be able to and to be allowed to;

a request, offer, suggestion, possibility, where **could** is more polite

may expresses the possibility and permission to do sth; a polite suggestion

might expresses a possibility (less possible than **may**) and a *hesitant* offer

must expresses a force, necessity, an *assumption*, an advice, a recommendation;

but **must not** expresses *prohibition* (!)

need not expresses that there is no necessity to do sth

shall expresses a suggestion

ought to and **should** express an advice, an obligation

will expresses a wish/request/demand/order (less polite than would); a prediction/assumption, promise, spontaneous decision, *habits*

would expresses a wish/request (more polite than will), habits in the past

Glossary

hesitant	unable to make a decision quickly
assumption	here a belief that sth is true
prohibition	a law or order that forbids sth
habit	a usual behavior

7.8 Grammar: Modal Auxiliaries

Task 1. *Fill the gaps with modals. Several modals may apply, depending on the intention you want to express. Remember to use the passive voice when necessary.*

The term smart .... (apply) to rather sophisticated systems.

Viscosity .... (change) when applying an electric or magnetic field.

Materials .... (make) that bend, expand or contract when a voltage is applied.

Recyclable materials .... further .... (develop).

Materials for more efficient fuel cells .... still .... (find).

Nanotubes .... (be) applicable in many ways.

The ecological impact of manufacturing materials .... (consider).

KEY

Chapter 1 Introduction

1.1 Historical Background

Task 1.

housing; communication; manipulate; clay, wood, skin; pottery; metals; heat; substances; alloy; properties; specimen; crystals; structure; properties; characteristics; technologically

Task 2.

to adjust: to change sth slightly to improve sth or make it more suitable for a particular purpose

to alter: to change sth so that it is better/more suitable

to change: to make sb/sth different

to modify: to make small changes to sth

to transform: to change sth completely

to vary: to make sth different

Task 3. Suggested Solution

X-ray diffraction is a powerful and readily available method for determining atomic arrangements in matter.

TEM is a microscope that produces an image by using electron beams that pass through a specimen, thus making possible the examination of internal features of the specimen.

SEM is a microscope that produces an image by using an electron beam that scans the surface of a specimen; an image is produced by reflecting the electron beam, thus making possible the examination of the surface of the specimen.

1.2 Grammar: Simple Past versus Present Perfect

Task 2.

have always played; were designated; had access; was first revealed; developed; have made; have been developed; have long envied; were not successful; mixed

1.3 Materials Science versus Materials Engineering

Task 1.

False. They develop/synthesize new materials.
False. New products are based on new and existing materials.
True.
False. This is done by materials scientists.

Chapter 1 Introduction – KEY 85

1.4 Selection of Materials

Task 1. Suggested Solution

Prohibitively is an adverb that modifies the adjective 'expensive'.

Task 2.

The characterization of in-service conditions.

The consideration of potential environmental deterioration.

Taking economics into account.

The trade-off of one property for another.

The development of a new material if in-service properties cannot be achieved.

A trade-off for economic reasons.

1.6 Case Study: The Turbofan Aero Engine

Task 1. Presentation Model (Figures mentioned in the text are not shown)

Good morning, Ladies and Gentlemen,

The topic of my presentation today will be the Turbofan Aero Engine. We will address three aspects; I will begin with the way the engine works. Then two materials that can be used for the engine's blades will be discussed, first a titanium alloy and next a carbon fiber reinforced polymer (CFRP).

As is well known, a turbofan engine is the most modern variation of a gas turbine engine. It is widely used to power large planes because of the high thrust it generates and its fuel efficiency.

This aerodynamic thrust is provided by air propelled through the engine by the turbofan. As can be seen from Figure 1, the air is compressed in the compressor, then mixed with fuel and burnt in the combustion chamber. The expanding gases drive the turbine blades and finally pass out of the rear of the engine, adding to the thrust.

Having briefly shown how a turbofan engine works, we will now look at two kinds of material which can be used for the blades, a titanium alloy and a carbon fiber reinforced polymer (CFRP).

As can be seen from the next slide, the alloy can be considered as a suitable material because of its sufficient hardness, high tensile strength and fatigue strength, the values of which are shown in this table in Figure 2.

As to in-service requirements, the alloy's excellent resistance to surface wear and corrosion would make it a material of choice. Its high density, on the other hand, leads to a reduction of payload, which clearly counts as disadvantageous.

Let's now address the second kind of material mentioned above, namely carbon fiber reinforced polymers. As the graph in Figure 3 shows, CFRP exhibits much lower density than the alloy and allows for more payload. On the other hand, the material's low toughness can lead to damage of the blades resulting from, e.g. bird strike.

To conclude, we can see that basically both materials discussed above can be applied in the construction of the turbofan's blades. But future development will probably show the replacement of weighty alloys by CFRP to allow for maximum payload. The problem of potential damage of blades from CFRP can be overcome by cladding, which means giving the material a metallic leading edge.

Thank you for your attention, Ladies and Gentlemen. I'll be pleased to answer questions now.

Chapter 2 Characteristics of Materials

2.1 Structure

Task. 1

subatomic: involves electrons in atoms and interaction with their nuclei

atomic: involves the organization of atoms relative to one another

nanoscopic: comprises molecules/atomic organizations within 1 nm – 100 nm

microscopic: can be viewed using a microscope

macroscopic: can be viewed with the naked eye

Task 2.

molecule; atom; electron; nucleus; proton; neutron

2.3 Case Study: The Gecko

Task 1.

underside; design; vertical; horizontal; sticky; mass; microscopically; molecules; surfaces; release; self-cleaning; adhesion; adhesive; residue

2.4 Property

Task 1.

It is a material trait that describes the kind and magnitude of response to a specific stimulus.

Yes, they relate deformation to an applied load or force; examples include elastic modulus and strength.

The thermal behavior of solids can be described by heat capacity and thermal conductivity.

Task 2. Suggested Solution

The graph in the figure above shows the relationship between the speed of crack propagation for metal in seawater at an increasing load. There are two curves showing the propagation of cracks with and without heat treatment before being exposed to an increasing load. Both curves rise steeply at first, the 'as-is curve' reaches 10^{-8} m/s, then flattens out. The curve representing the propagation of cracks of the heat-treated sample flattens out at about 10^{-9} m/s.

Chapter 2 Characteristics of Materials – KEY 87

2.5 Some Phrases for Describing Figures, Diagrams and for Reading Formulas

Task 1.

10,000	ten thousand
0.28	naught (zero) point two eight
$1/4$	a quarter, a (one) fourth
$1/12$	one over twelve
$6\,3/5$	six and three fifths
x^2	x squared
x^3	x cubed
x^{-4}	x to the power/part/order of minus four
$\sqrt{4}$	square root of four
$\sqrt[3]{a}$	cube root of a
$1/x$	reciprocal of x
a_n	a subscript (sub) n
$^n a$	a pre-superscript n

2.6 Grammar: Comparison

Task 1.

better; best. worse; worst. farther; farthest. further; furthest. less; least. smaller; smallest. more; most

2.7 Processing and Performance

Task 1. Suggested Solution

aluminum oxide; a printed page; is transparent; in the center is; reflected light is transmitted; opaque; light; the structure; single crystal; transparency; light reflected from the printed page.

2.8 Classification of Materials

Task 1.

True.

False. Ceramics will be used for electronics if their superconductivity at ambient temperatures is achieved.

False. Natural materials like wood or leather offer properties that, even with the innovations of today's materials scientists, are hard to beat.

2.9 Grammar: Verbs, Adjectives, and Nouns followed by Prepositions

Task 1. Suggested Solutions

Verbs: to rely on teamwork; to trade one property for another; the symbols relate to geometry.
Adjectives/Participles: a specimen transparent to light; results based on experiments; composites composed of two or more individual materials; according to this research paper
Nouns: in response to an electrical current; a decrease in temperature; in reference to Callister

Chapter 3 Metals

3.1 Introduction

Task 1. No Solution Suggested

Task 2. Suggested Solutions

copper Cu: A reddish-brown, ductile, malleable, metallic element; excellent conductor of heat and electricity; widely used for electrical wiring, water piping, and corrosion-resistant parts, either pure or in alloys such as brass and bronze.

nickel Ni: A silvery, hard, ductile, ferromagnetic metallic element; occurs in ores along with iron or magnesium; resists oxidation and corrosion; used to make alloys such as stainless steel; used in batteries, for electroplating and as a coating for other metals.

mercury Hg: A silvery white poisonous metallic element, liquid at room temperature; used in thermometers, barometers, vapor lamps and batteries and in the preparation of chemical pesticides; also called quicksilver.

sodium Na: A soft, light, extremely malleable silver-white metallic element; reacts explosively with water; naturally abundant in combined forms, especially in common salt; used in the production of industrially important compounds.

zinc Zn: A bluish-white, lustrous metallic element; brittle at room temperature but malleable when heated; used to form alloys including brass, bronze, various solders and nickel silver, to galvanize iron and other metals; used to make electric fuses; to produce roofing, gutters, and household objects.

aluminum Al: A silvery white, ductile metallic element, the most abundant in the earth's crust but found only in combination, chiefly in bauxite; good conductive and thermal properties; used to form many hard, light, corrosion-resistant alloys

gold Au: A soft, yellow, corrosion-resistant element, the most malleable and ductile metal; good thermal and electrical conductor; alloyed to increase its strength; used as an international monetary standard, in jewelry, for decoration, as a plated coating for electrical and mechanical components.

lead Pb: A soft, malleable, ductile, bluish-white, dense metallic element; used in containers and pipes.

tin Sn: A malleable, silvery metallic element; used to coat other metals to prevent corrosion; part of numerous alloys, such as pewter and bronze.

Chapter 3 Metals – KEY

3.2 Mechanical Properties of Metals

Task 1. Suggested Solution

Testing the tensile strength of a specimen is done by fixing one end and pulling at the other end (applying tensile load/stress), thereby causing tensile strain in the material, which means it gets longer and thinner.

Task 2

high; low. brittle; little. versus; elasticity

3.3 Important Properties for Manufacturing

Task 1.

ductility; copper; tin; gold; silver.

elasticity.

hardness; carbon.

malleability; lead.

Task 2 Suggested Solution

Durability

Durable metals do not corrode easily in air and damp conditions. Chromium and platinum have high durability but they are both expensive and in short supply. Gold, silver and aluminum are very durable too. We have to coat them against corrosion, e.g. by painting. Some fine steel alloys are very durable. One example is tungsten steel which is made of tungsten, chromium, carbon and iron.

3.4 Metal Alloys

Task 1. Suggested Solutions

How are metal alloys made?

Metal with metal alloys are made by mixing the molten substances and then cooling them until they solidify.

Which are the most often used alloys?

Common alloys are brass (copper and zinc) and aluminum alloys and steel.

What is the main alloying element in stainless steel?

Stainless steel contains chromium as predominant alloying element

Why are ferrous alloys produced in large quantities?

Iron is readily available since iron containing compounds are found in large quantities in the earth's crust.

What are ferrous alloys susceptible to?

They are inherently susceptible to corrosion in some common environments.

Why can ferrous alloys be described as versatile?

They can have a wide range of physical and mechanical properties.

3.5 Case Study: Euro Coins

Task 1. Suggested Solution

Colour; Electrical Conductivity; Malleability; Hardness; Corrosion Resistance; Recyclability; Antibacterial Characteristics

3.6 Grammar: Adverbs I

Task 1.

-ly; slowly. -ly; possibly. -ily; stickily. -ically; magnetically.

well. hard (hardly means almost not/none). fast. in a friendly way

Task 2.

directly attributable; extremely good; especially important; relatively economical; comparatively low ductility

Adverbs have to be used when one adjective modifies another.

3.7 Case Study: The Titanic

Task 1.

sulfur; phosphorous; manganese sulfide

Task 2

False. There were not enough lifeboats.

False. Not in 1912.

False. Sonar imaging revealed a series of six thin cuts, the damage totaling only 12 square ft, about the size of a human body.

True. The steel was full of large MnS (manganese sulfide) impurities that created weak areas and caused the metal to be brittle.

3.8 Grammar: The Passive Voice

Task 1.

is being printed; was edited; has been published; had been finished; will be copied; will have been handed in; would be done; would have been written

3.9 Case Study: The Steel-Making Process

Task 1.

called; be described; mixed; given; is produced; is converted; is removed; is added; used; be produced; shaped; be moved; molten; cooled; be lowered; dissolved; are oxidized

Chapter 4 Ceramics

4.1 Introduction

Task 1. Suggested Solution

Das griechische Wort *keramikos* zeigt, dass die erwünschten physikalischen Eigenschaften dieser Werkstoffe im Normalfall durch eine Wärmebehandlung bei hoher Temperatur erzielt werden, die man Brennen nennt.

Herkömmliche Keramiken bestehen aus dem Rohstoff Ton.

4.2 Structure of Ceramics

Task 1.

False. They are compounds between metallic and non-metallic elements.

False. Ceramics are compounds of at least two and often more elements.

4.3 Word Formation: Suffixes in Verbs, Nouns and Adjectives

Task 1. Suggested Collocations

suffix   example

suffix	example
-able/ -ible	a desirable property susceptible to fracture
-al	electrical conductivity
-ance -ence	resistance to corrosion in reference to a publication
-ant -ent	significant progress translucent ceramics
-ary	primary raw material
-ate	appropriate processing
-en	harden a metal
-ic/-ical	magnetic behavior a chemical reaction
-ify	identify cracks
-ion -ition	environmental pollution disposition to catastrophic failure
-ist	a renowned scientist
-ity	lack of ductility
-ive	corrosive environment

-ize -ization	characterize a specimen organization of elements in the periodic table
-ment	heat treatment
-ness	hardness of steel
-ous	porous ceramics
-ship	relationship between atomic structure and property

Task 2.

noun	adjective	verb
atom	atomic	to atomize
application	applicable	to apply for/to
bond, bonding	bonding	to bond
configuration	n.a.	to configure
dependence	dependent on	to depend on
example	exemplary	to exemplify
geometry	geometrical	n.a.
interaction	interactive	to interact with
notice	noticeable	to notice
softness	soft	to soften
solid, solidity	solid	to solidify
structure	structural	to structure
variable variability variance variant variation variety	various variable	to vary

4.4 Properties of Ceramics

Task 1.

strength; ductility; fracture; deformation; load; conductivities; characteristics; magnetic

Task 2.

transparent; transmitting light, one can see images clearly through the material
translucent; transmitting light but one cannot see images clearly because of diffusion
opaque; not transmitting any light, one cannot see through the material

Chapter 4 Ceramics – KEY 93

Task 3.
Applicability. The ceramics' applicability depends on their properties.

Susceptibility. One of the drawbacks of ceramics is their susceptibility to fracture.

Insulation. Ceramics show good insulation against heat and electricity.

4.5 Case Study: Optical Fibers versus Copper Cables

Task 1. Suggested Solution
Pros:

Glass fibers carry signals more efficiently and in a safer way than copper cables. They use less electrical power to transport the signals than copper cables. Due to their reduced weight and smaller width, optical fibers do not take up as much space as copper cables in underground cabling ducts.

Cons:

On the other hand, optical fibers are costlier than copper cables because they require sophisticated processing. They are more difficult to splice than copper cables which requires special training, which also makes them more expensive.

4.6 Grammar: Adverbs II

Task 1. Suggested Solution
Consequently, a new generation of these materials has evolved (consequently modifies the sentence).

The compounds are most frequently oxides, nitrides and carbides (frequently modifies the sentence).

Their crystal structures are generally more complex. (generally modifies the sentence).

Ceramics are relatively stiff and strong (relatively modifies the following adjectives).

Ceramics are typically very hard, extremely brittle, highly susceptible to fracture, typically insulative (the adverbs modify the following adjectives).

It is made of extremely high-purity silica (extremely modifies the following adjective).

They can't be spliced easily (the adverb modify the verb).

4.7 Case Study: Pyrocerams

Task 1.
Properties; Manufacturing; Applications

Task 2.
Thermal shock resistance is better than that of most other ceramics.

Pyrocerams exhibit excellent resistance to corrosion, too.

They can be formed by hot pressing fine powders and sintering them.

If constituents are varied, properties of the final ceramics vary too.

Sialons are the best materials for high-temperature use.

Typical uses of high-temperature ceramics include tubes for nonferrous metal melting.

4.8 Case Study: Spheres Transporting Vaccines

Task 1. Suggested Solutions

Researchers plan to enclose vaccines in spheres that are made from calcium phosphate, which is the main mineral constituent of bone. Knowing how the body dissolves bone fragments helped researchers to develop spheres that can be dissolved by osteoclasts, thus releasing the vaccine.

The same way as osteoclasts attack stray pieces of bone in the body and dissolve them, they can dissolve the calcium phosphate of the spheres so that the vaccines can be delivered to the body.

Vaccines have to be kept dry in order not to deteriorate or break down before release.

4.9 Useful Expressions for Shapes and Solids

Task 1. Suggested Solutions

circular; a completely round, flat shape

conical; a solid/hollow object with a round flat base and sides that slope up to a point

cubic; a solid/hollow figure with six equal square sides

cylindrical; a solid/hollow figure with round ends and long straight sides

a thin flat round object

elliptic; a regular oval shape

hemispherical; one half of a sphere

hexagonal; a flat shape with six straight sides and six angles

pentagonal; a flat shape with five straight sides and five angles

prismatic; a solid with parallel same-size ends and with sides whose opposite edges are equal and parallel

rectangular; a flat shape with four equal sides and four angles

rhomboid; a flat shape with four equal sides and four angles which are not 90°

a half circle

spherical: a solid figure that is completely round with every point on its surface at an equal distance from the center

square; a shape with four equal sides and four angles of 90°

star-shaped; resembling a star

trapezoid; a shape with four straight sides, none of which are parallel

triangular; a flat shape with three straight sides and three angles

Chapter 5 Polymers

5.1 Introduction

Task 1.

rubber; cotton; silk; wool; leather; plants; animals; molecules; synthetic; industry; properties; produced; applications

Task 2. Diagram Model

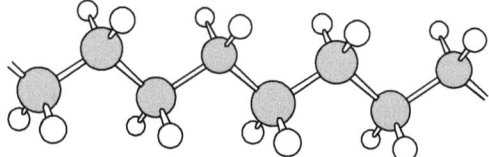

Figure 19:
Schematic structure of polyethylene

5.2 Word Formation: The Suffix -able/-ible

Task 1. Suggested Solutions

-able:	appreciate	appreciable elongation of a specimen
	attribute	properties attributable to atomic structure
	compare	comparable data
	desire	desirable magnetic properties
	notice	noticeable stretching
	rely	reliable resources
	suit	a suitable biopolymer
-ible:	access	easily accessible information
	flex	flexible rubbers
	perceive	perceptible changes in temperature
	reproduce	reproducible results

5.3 Properties of Polymers

Task 1.

polyethylene; polyvinylchloride; polycarbonate; polystyrene

Task 2. Suggested Solutions

This means they tend to soften and/or decompose at modest temperatures in some instances.

Polymers typically have low densities. That's why they are not as stiff or as strong as ceramics or metals.

Polymers are extremely ductile and pliable. This means they are easily formed into complex shapes.

In general they are relatively inert chemically. That's why they can be used in a large number of environments.

Polymers have low electrical conductivities and are nonmagnetic. This property may prove to be advantageous in some applications.

5.4 Case Study: Common Objects Made of Polymers

Task 1. Suggested Solutions

object	property
billiard balls	high density, hard, not elastic
bike helmets	hard, ductile, low weight, has to absorb energy
plastic knives	stiff, not too brittle, chemically inert, non-toxic, recyclable
water bottles	inert, low weight, good plastic deformation, non-toxic, recyclable

5.5 Case Study: Ubiquitous Plastics

Task 1.

commodity	merchandise
cushion	soft, protective pad
foam	bubbles of air together in a mass
luminescent	emitting light
payload	total weight that an airplane can carry
spongy	resembling an artificial or natural material that is soft, light and full of holes
ubiquitous	found everywhere

Task 2. Suggested Solutions

architectural design: The Allianz Arena has a plastic facade with a film skin which allows for innovative design and special effects.

interior design: PU mats at a Prada store have a textured surface, are luminescent and can be cut out for displays.

aircraft engineering: Airplanes are engineered with fiber composites which reduce weight, leading to less fuel consumption and increased payload and range.

5.6 Grammar: Reported Speech (Indirect Speech)

Task 1.

knew; had read; had been; had been interested; would know

Task 2.

Dirk Ziem confirmed that plastics nowadays had a high quality image, that the elegant appearance of the iPod could not be topped and the functionality of modern athletic wear would not be surpassed soon.

According to the *San Francisco Chronicle*, plastic furniture had become the focal point in some of the most elegantly designed rooms.

Rem Kohlhaas stated that no other materials could be so lightweight and luminescent.

5.7 Polymer Processing

Task 1. Diagram Model

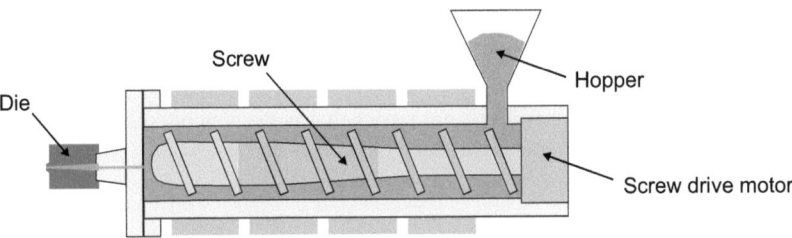

Figure 20: Schematic drawing of an extruder

Task 2.

manufacture; pressure; solidified; is ejected; metal; shaped; articles

Task 3. Suggested Solution

Blow molding is used to produce plastic containers and bottles. The polymer is melted and extruded into a hollow tube. Then the semi-molten polymer is placed in a cooled metal mold. Air or steam is then blown under pressure into the tube so that the tube walls fit to the contours of the mold; thus the tube is shaped into a hollow bottle or container. After cooling sufficiently, the mold is opened and the part is ejected.

5.8 Case Study: Different Containers for Carbonated Beverages

Task 1. Suggested Solution

As manufacturers we choose plastic as material for our soft drink bottles, since it is inexpensive and can be recycled. Labels can easily be glued on. The fact that plastic is a poor barrier to the diffusion of CO_2 is of no consequence since the bottles we produce are rather small. Unfortunately, beer does not sell well in plastic bottles, that's why we use cans and/or glass bottles for beer. In general, we are aware of the hazards posed by broken glass and we would gladly do without glass bottles if consumers accepted higher prices, as cans are relatively expensive to produce.

As consumers we prefer aluminum alloy for beer cans as it is relatively strong, cools rapidly and keeps carbonization well. For soft drinks we'd rather choose small plastic bottles because they are lightweight and the top can be screwed tight. We usually buy soda water in large glass bottles since they can be delivered to the home, are recyclable and the glass is a good barrier to the diffusion of CO_2.

Chapter 6 Composites

6.1 Introduction

Task 1.

wood; cellulose; bone; aerospace; underwater; transportation; corrosion; artificially; dissimilar; phases

Task 2.

2 or more.

The design goal is to achieve a combination of properties that is not exhibited by any single material, and also to include the best properties of each of these materials.

6.2 Case Study: Snow Ski

Task 1. Suggested Solution

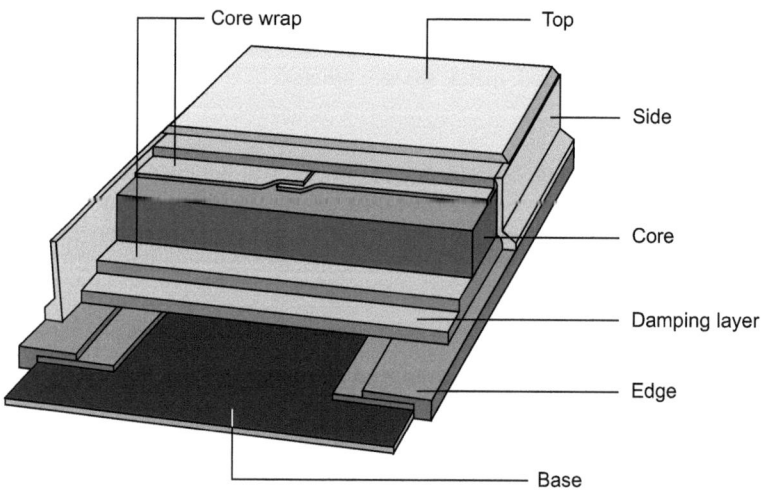

Figure 21: Cross-section of a snow ski

Chapter 6 Composites – KEY

Task 2. Suggested Solutions

provide:	Ceramics can provide insulation.
be used for:	Aluminum is used for light-weight constructions.
act as:	Graphite can act as lubricator.
function as:	Composites function as structural materials in aerospace engineering.
facilitate:	The lotus effect facilitates keeping surfaces clean.
improve:	Fiber reinforcement improves tensile strength.

6.3 Grammar: Gerund (-ing Form)

Task 1.

avoiding; freezing; stopping

Task 2.

In spite of studying hard, he didn't pass the exam.

He looked forward to finishing the academic year.

The edge of the ski facilitates turning by cutting into the snow.

She saw the new instrument on entering the lab.

He started the instrument without reading the manual first.

Task 3.

good/bad at working; angry about losing; disappointed about seeing; famous for giving; interested in hiring

the advantage of using; the danger of being exposed

accused of plagiarizing; concentrates on developing; cope with solving; get used to working; decided against using

Task 4.

includes writing; justify paying; avoid exposing; considered varying; admits having missed; suggests working

6.4 Case Study: Polymer-Matrix Composites (PMCs)

Task 1. Suggested Solution

CFRPs are applied in, e.g. racecar manufacturing for the chassis as well as other components of high-end race cars. Since low weight is essential, the material is used for its excellent strength-to-weight ratio in spite of the high cost of manufacturing carbon fibers. Recently, manufacturers have also started to use CFRP for body panels in everyday road cars because of its increased strength and decreased weight, which results in lower fuel consumption.

6.5 Word Formation: Prefixes

Task 1. Suggested Solutions

a-	*atypical behavior*
aero-	aerospace engineering
anti-	antibacterial coating
auto-	autofocus camera
bi-	bifunctional crosslinking
bio-	biocompatible material
co-	coauthor of a publication
counter-	corrective countermeasures
de-	deactivating agent
di-	dielectric constant
dis-	dissimilar approaches
eco-	ecofriendly energy
ex-	exchange program
geo-	geometrical figure
im-	impurity atoms
inter-	interatomic bonding
kilo-	kilometer per hour
macro-	macromolecular chain
mal-	malfunction indicator
mega-	megawatt electricity-generating unit
micro-	microstructured fiber
milli-	milligrams per cubic meter
mis-	miscalculate effects
multi-	multicolored spheres
nano-	carbon nanotubes
non-	non-linear algebra
out-	outstanding performance
over-	overdue report
poly-	polyethylene bags
pre-	prefab building material

proto-	prototype testing
re-	recrystallization temperature
semi-	semiconductor chip
sub-	subzero temperature
super-	supercooled liquid
trans-	transformation mathematics
tri-	triangular shape
ultra-	ultralight aircraft
under-	underestimated value
uni-	unidirectional flow

Chapter 7 Advanced Materials

7.1 Introduction

Task 1. Individual solutions possible.

7.2 Semiconductors

Task 1.

Si silicon; Ge germanium

Task 2. Suggested Solutions

Do semiconductors have high electrical conductivity? Are semiconductors good electrical conductors?	Semiconductors have electrical properties that are intermediate between the electrical conductors.
What are the electrical characteristics of these materials sensitive to? What do the electrical characteristics of these materials depend on?	The electrical characteristics of these materials are sensitive to the presence of impurity atoms.
When does the intrinsic electron and hole concentration increase? What effect do rising temperatures have on hole concentration?	The intrinsic electron and hole concentration increases with rising temperatures.
How are semiconductors classified? What makes a semiconductor intrinsic or extrinsic?	Semiconductors are classified as either intrinsic or extrinsic on the basis of their electrical behavior.

7.3 Case Study: Integrated Circuits

Task 1.

miniaturized; manufactured; electronic; perform; advancements; improvement; approach; consume

7.4 Grammar: Subordinate Clauses

Task 1.

Scientists introduced semiconductors in order to make possible the development of integrated circuitry.

The audience stayed in the lecture hall so that they could hear the second lecture.

Researchers added impurities so as to optimize conductivity.

For the system not to overload, circuit breakers were installed.

7.5 Smart Materials

Task 1

Smart materials are able to sense changes in their environments and then respond to these changes or stimuli in a predetermined manner. These traits are also found in living organisms. Smart materials have a sensor that detects an input signal and an actuator that triggers a response and adaptation. Actuators can initiate changes of shape, position or mechanical characteristics in response to changes in temperature, pressure, light, electric fields and/or magnetic fields.

Task 2.

Shape Memory Alloys

… where deformation can be caused through temperature changes.

… their original shapes when the temperature is changed.

Piezoelectric Ceramics

… an electric field when their dimensions are altered.

Magnetostrictive Materials

… they respond to magnetic fields.

Electrorheological/Magnetorheological Fluids

… by applying an electric or magnetic field.

7.6 Nanotechnology

Task 1. Suggested Solution

In the so-called top-down approach to the chemistry and physics of materials, researchers study large and complex structures first and then investigate smaller fundamental building blocks of these structures.

Chapter 7 Advanced Materials – KEY

In the so-called bottom-up approach, researchers arrange atoms in order to develop and study mechanical, magnetic and other properties which would otherwise not be possible.

7.7 Case Study: Carbon Nanotubes

Task 1.

consists; thickness; end; molecule; atoms; diameter; ductile; known; efficient; applicable; fields

7.8 Grammar: Modal Auxiliaries

Task 1. Suggested Solution

can be applied; will change; could be made; ought to be further developed; might still be found; can be ; must be considered

Credits

The author acknowledges the following copyright owners and wishes to thank for the kind permission to use their materials.

Michel F. Ashby and David R. H. Jones, *Engineering Materials 1*, Excerpts of pp 3–322, Copyright Elsevier, 3e 2005

William D. Callister Jr., *Materials Science and Engineering: An Introduction*, Excerpts of pp 2–579, Copyright John Wiley & Sons 7e 2007. Reproduced with permission of John Wiley & Sons, Inc.

Ram Seshadri, Class Materials 100A, UC Santa Barbara, Engineering, Fall 2007

Vision Works. The fascinating world of polymers Issue 1/2006, "Plastics are the future, my boy. From cheap substitute to natural material of the future". Excerpts of pp14–17, Copyright Bayer MaterialScience AG, Leverkusen, 2006

Every effort has been made to identify the sources of all the materials used or to trace right holders, but I apologize if any have inadvertently been overlooked.

Selected Reference List

Celeste Biever, "What Vaccine Design Can Take from Bones", *New Scientist*, March 18 2006, Volume 189, No 2543

Peter A. Thrower, *Materials in Today's World*, The McGraw-Hill Companies, Inc. 2e 1996

Läpple, V.: *Einführung in die Festigkeitslehre*. Wiesbaden: Vieweg+Teubner, 2008

Trzesniowski, M.: *Rennwagentechnik*. Wiesbaden: Vieweg+Teubner, 2010

Wikipedia:

http://commons.wikimedia.org/wiki/File:Materials_science_tetrahedron;structure,_processing,_performance,_and_properties.JPG
December 28, 2010

http://en.wikipedia.org/wiki/Turbofan
December 28, 2010

http://en.wikipedia.org/wiki/Titanic
December 28, 2010

http://en.wikipedia.org/wiki/Gießen_(Verfahren)
January 7, 2011

Dictionaries Recommended for Students (the latest available edition of the printed versions should be used)

Dictionary of Contemporary English For Advanced Learners. Langenscheidt Longman, Pearson Education Limited

Oxford Advanced Learner's Dictionary. Oxford University Press

Oxford Thesaurus of English. Oxford University Press

Oxford Collocations Dictionary for Students of English. Oxford University Press

Selected Reference List

The American Heritage Dictionary of the English Language.
Houghton Mifflin Company. Boston, New York, London
www.thefreedictionary.com
www.leo.org

Glossary

abrasion, to abrade	the process of being rubbed away by friction, to rub away
abrasive, *n*, *adj*	a substance that abrades, abrading
acid	a chemical, usually a sour liquid, that contains hydrogen with a pH of less than 7
adhesive *n*, *adj*, to adhere, adhesion, *n*	a substance used for joining surfaces together, sticky
alloy	a metallic substance that is composed of two or more elements which keep the same crystal structure in the alloy
ambient temperature	the temperature of the air above the ground in a particular place; usually room temperature, around 20 – 25 °C
aqueous	watery
assumption	here a belief that sth is true
axle	a supporting shaft on which wheels turn
bearing	a device to reduce friction between a rotating staff and a part that is not moving
binder	a polymeric material used as matrix in which particles are evenly distributed
blast furnace	the oven in which ore is melted to gain metal
boundary	the interface separating two neighboring regions having different crystallographic orientation
brick	a rectangular block of baked clay used for building
china	high-quality porcelain, originally made in China
to clad	to cover a material with a metal
clay	a kind of earth that is soft when wet and hard when dry
combustion	the process of burning; here of fuel
commodity	article of trade
compound	a pure, macroscopically homogeneous substance consisting of atoms/ions of two/more different elements that cannot be separated by physical means
conductivity	ability to transmit heat and/or electricity
constituent phase	one of the phases from which a substance is formed
corrosive, *n*, *adj* to corrode, corrosion	a corroding substance, e.g. an acid
to counterfeit	to make a copy of sth with criminal intent, to fake
counterpart	here sth that has a similar function
crack, *n*, *v*	a break, fissure on a surface
creep, *n*	time-dependent permanent deformation of materials at high temperatures or stress
to cure	to improve the properties of polymers and rubber by combining with, e.g. sulfur under heat and pressure; cf. to vulcanize
to damp(en)	to make sth less strong, to soften
to decompose	to change chemically, to decay
denomination	a unit of value, esp. for money

Glossary

dense, density, *n*	referring to mass per volume
density	mass per volume
to derive	to deduce; to obtain (a function) by differentiation
die	here a metal block containing small holes through which the polymer is forced
dielectric constant	a measure of a material's ability to resist the formation of an electric field within it
diffusion	the movement of atoms/molecules from an area of higher concentration to an area of lower concentration
to disperse, dispersion, *n*	to distribute particles evenly through a medium
disposition	a physical property/tendency
duct	a pipe for electrical cables and wires
duct tape	an adhesive tape for sealing heating and air-conditioning ducts
ductility, *n* ductile, *adj*	a material's ability to suffer measurable plastic deformation before fracture
elastic modulus (E)	or Young's Modulus, a material's property that relates strain (ε, epsilon) to applied stress (σ, sigma)
to etch	to cut into a surface, e.g. glass, using an acid
fatigue	the weakening/failure of a material resulting from prolonged stress
ferrous	of or containing iron
flammable	easily ignited, capable of burning, inflammable
fracture toughness	the measure of a material's resistance to fracture when a crack occurs
fullerene	carbon molecule named after R. Buckminster Fuller, sometimes called buckyball, composed entirely of C in the form of a hollow sphere, ellipsoid or tube
glass transition temperature T_g	the temperature at which, upon cooling, a non-crystalline ceramic or polymer transforms from a supercooled liquid to a solid glass
grain boundary	a line separating differently oriented crystals in a polycrystal
habit	a usual behavior
hesitant	unable to make a decision quickly
hull	the body of a ship
to ignite, ignition, *n*	to begin to burn, to cause to burn
impact	a high force or load acting over a short time only
impurity atoms	here atoms of a substance that are present in a different substance
in	inch, 2.54 cm
integrated circuit	millions of electronic circuit elements incorporated on a very small silicon chip
interface	the area between systems where they come into contact with each other
lb	pound, 453.592 grams
lustrous, luster, *n*	shining brightly and gently

malleability	the property of sth that can be worked/hammered/shaped without breaking
manual assembly	putting together manufactured parts to make a completed product by hand
matrix	a substance in which another substance is contained
median	relating to or constituting the middle value in a distribution, e.g. the median value of 17, 20 and 36 is 20
minute	extremely small
monomer	a molecule that can combine with others of the same kind to form a polymer
nm	nanometer (10^{-9} m)
nozzle	a device with an opening for directing the flow of a liquid
ore	a mineral from which a metal can be extracted
pear-shaped	having a round shape becoming gradually narrower at the end
perpendicular to	forming an angle of 90° with another line/surface
phase	a form or state of matter (solid/liquid/gas/plasma) depending on temperature and pressure
phenomenon, phenomena, *pl*	a fact/event that can be identified by the senses
pig iron	crude iron
plastic deformation	a non-reversible type of deformation, i.e. the material will not return to its original shape
predetermined	decided beforehand
prohibition	a law or order that forbids sth
propagation	the process of spreading to a larger area
to refine	to make/become free from impurities
reflectivity	the ability to reflect, i.e. to change the direction of a light beam at the interface between two media
refraction	the bending of a light beam upon passing from one medium into another
release, *v*, *n*	to let go
residue	the remainder of sth after removing a part
resilience, *n* resilient, *adj*	elasticity; property of a material to resume its original shape/position after being bent/stretched/compressed
resin	a natural substance, e.g. amber, or a synthetic compound, which begins in a highly viscous state and hardens when treated
resistivity	a material's ability to oppose the flow of an electric current
rocketry	the science and technology of rocket design, construction and flight
rod	a thin, straight piece/bar, e.g. of metal, often having a particular function
scanning probe microscope	(SPM), a microscope that scans across the specimen surface line by line, from which a topographical map of the specimen surface (on a nanometer scale) is produced
to scatter	to distribute in all directions
scrap iron	metal objects that have been used
to shatter	to break suddenly into very small pieces

slip casting	the process of pouring liquefied material into a mold; after the liquid is drawn out, the solid is removed from the mold
slope	a line that moves away from horizontal
sonar	a system using transmitted and reflected underwater sound waves to detect/locate/examine submerged objects
sphere	a solid figure that is completely round
to splice, e.g. cables	to join two pieces at the end
starch	a white, tasteless powder found in plants, e.g. rice, potatoes
strain	the response of a material when tensile stress is applied
to stray	to move away from the place where sth/sb should be
strength	the power to resist stress or strain; the maximum load, i.e. the applied force, a ductile material can withstand without permanent deformation
stress, n	the force applied to a material per unit area; (σ, sigma = F/A or lb/in^2)
supercooled	cooled to below a phase transition temperature without the occurrence of transformation
to be susceptible to susceptibility, n	to be easily affected/influenced by
to synthesize, synthesis, n	to produce a substance by chemical or biological reactions
t/s	tons per second
to tap	to remove by using a device for controlling the flow of a liquid
to tarnish tarnish, n	to discolor a metal surface by oxidation, to become discolored
tensile stress	a force tending to tear a material apart
thermoplastic, n, adj	a polymer that softens when heated and hardens when cooled
thermoset	a polymeric material that, once having cured or hardened by chemical reaction, will not soften or melt when heated
thrust	a forward directed force
tile	a flat, square piece of material
toe pad	a cushion-like flesh on the underside of animals' toes and feet
torsion, torsional, adj	the stress/deformation caused when one end of an object is twisted in one direction and the other end is twisted in the opposite direction
tube	a long hollow pipe through which liquids/gases move
vacuum tube	an electron tube from which all or most of the gas has been removed, letting electrons move without interacting with remaining gas molecules
velvet	a type of cloth with a thick, soft surface
viscous, adj viscosity, n	having a relatively high resistance to flow
yield strength	the point at which a material starts to deform permanently
Young's Modulus	elastic modulus (E), a material's property that relates strain (ϵ, epsilon) to applied stress (σ, sigma)

n = noun adj = adjective v = verb

Aus dem Programm Werkstofftechnik

Rösler, Joachim / Harders, Harald / Bäker, Martin
Mechanisches Verhalten der Werkstoffe
3. durchges. u. korr. Aufl. 2008. XIV, 521 S. mit 319 Abb., 31 Tab. u. 34 Aufg. m. Lös. Br. EUR 32,90
ISBN 978-3-8351-0008-4

Bonnet, Martin
Kunststoffe in der Ingenieuranwendung
verstehen und zuverlässig auswählen
2009. XII, 282 S. mit 269 Abb. Br. EUR 24,90
ISBN 978-3-8348-0349-8

Weißbach, Wolfgang
Werkstoffkunde
Strukturen, Eigenschaften, Prüfung
17., akt. u. überarb. Aufl. 2010. XVI, 432 S. mit 298 Abb. und 246 Tab. Br. EUR 28,90
ISBN 978-3-8348-0739-7

Kuna, Meinhard
Numerische Beanspruchungsanalyse von Rissen
Finite Elemente in der Bruchmechanik
2., verb. Aufl. 2010. XII, 446 S. mit 277 Abb. und 11 Tab. Br. EUR 49,95
ISBN 978-3-8348-1006-9

Abraham-Lincoln-Straße 46
65189 Wiesbaden
Fax 0611.7878-400
www.viewegteubner.de

Stand Juli 2010.
Änderungen vorbehalten.
Erhältlich im Buchhandel oder im Verlag.

Aus dem Programm Maschinenelemente

Wittel, Herbert / Muhs, Dieter / Jannasch, Dieter / Voßiek, Joachim
Roloff/Matek Maschinenelemente
Normung, Berechnung, Gestaltung - Lehrbuch und Tabellenbuch
19., überarb. u. erw. Aufl. 2009. XX, 808 S. mit 711 Abb. 75 vollst. durchgerechn. Beispl. und einem Tabellenbuch mit 282 Tab. mit VIII, 240 S. Geb. mit CD EUR 36,90
ISBN 978-3-8348-0689-5

Wittel, Herbert / Muhs, Dieter / Jannasch, Dieter / Voßiek, Joachim
Roloff/Matek Maschinenelemente Formelsammlung
Interaktive Formelsammlung auf CD-ROM
10., überarb. Aufl. 2010. VIII, 296 S. Br. EUR 21,95
ISBN 978-3-8348-1328-2

Muhs, Dieter / Wittel, Herbert / Jannasch, Dieter / Voßiek, Joachim
Roloff/Matek Maschinenelemente Aufgabensammlung
Aufgaben, Lösungshinweise, Ergebnisse
15., überarb. u. erw. Aufl. 2010. VIII, 366 S. Br. EUR 26,95
ISBN 978-3-8348-1259-9

TEDATA (Hrsg.)
Roloff/Matek Bauteilkatalog
Maschinen- und Antriebselemente
Erzeugnisse und Hersteller nach eCl@ss, CD mit Zugangsdaten zur Bauteildatenbank online
2009. XVIII, 354 S. Br. EUR 29,90
ISBN 978-3-8348-0922-3

**VIEWEG+
TEUBNER**

Abraham-Lincoln-Straße 46
65189 Wiesbaden
Fax 0611.7878-400
www.viewegteubner.de

Stand Juli 2010.
Änderungen vorbehalten.
Erhältlich im Buchhandel oder im Verlag.

MIX
Papier aus verantwortungsvollen Quellen
Paper from responsible sources
FSC® C105338

If you have any concerns about our products,
you can contact us on
ProductSafety@springernature.com

In case Publisher is established outside the EU,
the EU authorized representative is:
**Springer Nature Customer Service Center GmbH
Europaplatz 3, 69115 Heidelberg, Germany**

Printed by Libri Plureos GmbH
in Hamburg, Germany